Lecture Notes in Mathematics

Edited by A. Dold and B. Eckmann

447

Sue Toledo

Tableau Systems for First Order Number Theory and Certain Higher Order Theories

Springer-Verlag
Berlin · Heidelberg · New York 1975

Dr. Sue Toledo
Dept. of Mathematical Statistics
Columbia University
in the City of New York
New York, NY 10027/USA

Library of Congress Cataloging in Publication Data

Toledo, Sue Ann, 1940-
Tableau systems for first order number theory and certain higher order theories.

(Lecture notes in mathematics ; 447)
Bibliography: p.
Includes index.
1. Proof theory. 2. Numbers, theory of. 3. Predicate calculus. I. Title. II. Series: Lecture notes in mathematics (Berlin) ; 447.
QA3.L28 no. 447 [QA9.54] 510'.8s [511'.3]
75-6738

AMS Subject Classifications (1970): 02-02, 02A05, 02B15, 02D99, 02H15

ISBN 3-540-07149-0 Springer-Verlag Berlin · Heidelberg · New York
ISBN 0-387-07149-0 Springer-Verlag New York · Heidelberg · Berlin

Offsetdruck: Julius Beltz, Hemsbach/Bergstr.

TABLE OF CONTENTS

INTRODUCTION

Hilbert's Program. In 1925 and 1927, in his papers "On the Infinite" [1] and "The Foundations of Mathematics" [2], Hilbert gave a comprehensive presentation of his ideas regarding the foundations of mathematics. Here he started from the point of view that for logical inference (and thus science in general) to be reliable, it had to be concerned with extra-logical concrete objects, which were to be given to the understanding through immediate intuitive experience prior to all thought. Furthermore, all the properties of the objects that could be meaningfully considered -- such as that they occurred, or differed, or followed each other -- had also to be immediately given intuitively. In particular, they would have to be decidable. Consequently, the completed infinite totality was something merely apparent, and for an existential statement to be meaningful, it would have to refer to an object that could actually be given in some way. Hilbert further believed that the paradoxes of set theory had come about due to the illegitimate use of arbitrary abstract notions, in particular those under which infinitely many objects were subsumed. On the other hand, it seemed that much of mathematics could not be carried out under such restrictions. Indeed, Hilbert felt, the classical modes of inference seemed to "mirror" our very thought processes. In such a situation, what was to be done?

Recalling how Weierstrass had provided a foundation for the infinitesimal calculus by reducing propositions about infinitesimals to relations between finite magnitudes, Hilbert suggested that the modes of inference still employing the infinite in mathematics could similarly be replaced by finite processes. But in this case the replacement was to be such that we were to be able to continue precisely as before, using the same classical modes of reasoning as always.

But how was this apparently self-contradictory goal to be achieved, a goal that in any case seemed unreasonable after the penetrating finitistic criticisms of classical mathematical methods that had been made by Brouwer and Hilbert, among others?

Hilbert made the following remarkable proposal. From Peano, Frege, Russell and Whitehead it was known that it was possible to formalize classical mathematics, thus transforming its theorems and proofs into finitary objects, mere strings of symbols put together in finitarily specified ways. Recalling the original controversy concerning the introduction of imaginary numbers to simplify the laws of existence and numbers of roots of an equation -- "how can one trust a proof that passes through reasoning about nonexistent objects" -- Hilbert saw the formalized infinite totalities and nonfinitary propositions as ideal elements that could be introduced into mathematics to simplify the laws of inference. If one could now give a finitary proof that this introduction of ideal elements could not yield an inconsistency, then one would have a finitary justification of the use of classical methods! More specifically, as Hilbert later pointed out as part of his response to Brouwer's objection that consistency wasn't sufficient (it was truth that was needed), a finitary consistency proof for a formal system would yield a finitary proof of any proposition proved within the system that had finitary content (these propositions being, according to Hilbert, the universal arithmetical formulas or formulas resulting from these when numerals are substituted for variables). This is because in such a case the consistency proof would make it impossible for any counterexample to the universal formula to be true, because, namely, then it would also be provable, yielding an inconsistency.

Responses to Hilbert's idea were varied. Most mathematicians with an interest in the problem probably joined Weyl in hoping for fruitful and illuminating consequences from Hilbert's suggestion, while questioning whether this was the most economical and meaningful approach to coming to an understanding of the meaning and role of finitism in mathematics. Brouwer, who was himself at the time trying to directly reconstruct mathematics on an "intuitionistic" basis, just could not consider reasonable an approach to mathematics that reduced it to a "formula game." (And he seemed to have been so forceful in his criticism as to actually bring Hilbert in 1927 to simultaneously defend the correctness and <u>meaningfulness</u> of the nonfinitary law of the excluded middle and to compare his approach to the widely accepted positivistic view of the role of physical theories according to which the only important thing about a theory is that the experimentally verifiable consequences check out. In physics also, Hilbert noted, the nonverifiable statements were often looked upon as <u>meaningless</u>.)

But the greatest challenge to Hilbert's program was soon to come: <u>Gödel's Incompleteness Theorems</u>. In 1930, Gödel's proof of the completeness of the first order predicate calculus [1] had lent Hilbert's formalistic conception support, indeed was a favorable resolution of Problem IV posed by Hilbert in 1928 [3]. Gödel, however, in fact totally disagreed with Hilbert (and Brouwer) on the most fundamental issue in question: Gödel did <u>not</u> believe that classical, nonfinitary, mathematics was meaningless, that meaning could be attributed only to propositions that speak of concrete and finite objects. (Cf.

Wang [1], pp. 8-11, for a letter in which Gödel attributes to his <u>non-finitary</u> viewpoint the fact that he obtained the completeness proof where others had failed. And this, he further says, also applies to his other work to a large extent.) In the 1928 paper just mentioned, Hilbert had also posed (as Problem III) the problem of proving the completeness of the axiom systems for number theory and analysis. In 1931 Gödel published two papers, [2], [3], in which he first resolved this problem <u>negatively</u>, and then went on to provide a result that implied that Hilbert was wrong in his most basic assumption and that his program could not be carried out as he had planned.

Gödel first developed a method which yielded (given Turing's work [1]) that mathematics cannot be formalized in any single formal system, that in particular, given any formal system containing first order arithmetic, it is possible to find a true statement with <u>finite content</u> (a universal statement) that can be neither proved nor refuted within the system. This meant that the part of Hilbert's program involving the formalization of mathematics was perhaps as much of a problem as finding a consistency proof in the sense that if one wished to formalize <u>all</u> of mathematics, one would have to continually provide new formal systems including axioms based on new ideas (as well as consistency proofs for each successive system, consistency proofs, indeed, that would also continually have to involve new ideas, as followed from Gödel's second incompleteness theorem -- cf. below). Since, however, almost all the mathematics that mathematicians were interested in <u>could</u> be formalized in one system, say Zermelo-Fraenkel set theory, this result, although surprising to Hilbert, did not constitute a real block to his program of using formalism to give a finitistic foundation to the mathematics that mathematicians were actually doing.

But Gödel's second incompleteness theorem dealt Hilbert's program a much more serious blow. This result was that one could never prove the consistency of a formal system for mathematics within the system itself. Since the kinds of formal systems that one needs to formalize modern mathematics (such as those that Hilbert considered) include all the finitary methods used in mathematics, this means that it is impossible to prove the consistency of a formal system for mathematics with finitary means. With this, Hilbert's grand idea designed to procure an absolutely unassailable foundation for classical mathematics was apparently completely vanquished.

Further Issues and New Possibilities. Gödel again. In 1933 Gödel published a result that showed that there were further things to be learned from this topic that it seemed he had just closed for good. In [4] Gödel showed how to transform any proof of a formula in first order arithmetic to an intuitionistic proof of a classically equivalent formula. (Gentzen had arrived at this result independently at about the same time.) This result showed, as Gödel pointed out, that intuitionistic number theory was only apparently weaker than classical number theory, for anything that could be proved classically could also be proved intuitionistically, although with a certain change in meaning. On the one hand this pointed out to those who held the view that finite, constructive, intuitionistic, and classical methods were just successive additions to one's store of tools (i.e. so that more and more could be proved -- a view probably held by most mathematicians) that in this case at least this wasn't so, that the issue was a truly philosophical one and lay at the level of fundamental differences in meaning.

On the other hand it suggested to the proof theorists both (1) that one might give up the view of classical inferences as being meaningless "ideal objects" one passed through, and look for a more fundamental reason for the fact that classical reasoning gave correct finitary theorems, e.g. based on this or another reinterpretation of classical reasoning; and (2) that one could perhaps obtain the desired consistency proofs if one settled for means less restrictive than finitary methods (even if intuitionistic methods were decided to be unacceptable).

It should be pointed out that at the time this result came to some as something of a shock. This was because there didn't exist then (nor is there to this day) any universally accepted precise formulation of the distinction between finitistic and intuitionistic views of truth, meaning, correct reasoning, etc. Indeed even the debates between Hilbert and Brouwer had failed to really address the issue and seemed rather to consist in arguments about the path to take to reach their common goal of founding mathematics on its only meaningful part, which part they both saw as dealing with objects that can be given by construction, a part in which the law of excluded middle could not be indiscriminately applied, etc. Thus Gödel had forced the proof theorists into a paradoxical situation: as long as no sharp distinction was made between "finitism" and "intuitionism," the 1931 incompleteness results seemed to imply that the desired consistency proof could not be given, while the 1932 result gave one.

Gentzen. In 1936 Gentzen [1] published a consistency proof for first order number theory that employed as its only noncombinatorial principle transfinite induction over all ordinals less than ϵ_0 , a mode of reasoning that he considered finitary, and which in any case seemed to most logicians to be constructive. In [3] Gentzen further showed directly that, although transfinite induction up to any ordinal strictly less than ϵ_0 could be proved in first order number theory, it was impossible to prove transfinite induction up to ϵ_0. (The first part of this had also been shown in Hilbert and Bernays [1]; the second, of course, also followed from the first by Gödel's result.) Thus Gentzen had apparently found more elementary means than those of classical number theory that could be used to prove its consistency. He had further isolated in transfinite induction up to a certain constructive ordinal a particular (and especially interesting because of its simplicity and connection with other foundational problems) noncombinatorially justified combinatorial principle that could be used to obtain a consistency proof.

The works of Gentzen along these lines were the basis of much of the subsequent work in proof theory, and it is to the description of many of the major contributions in this area that most of this volume is devoted (see below for details).

Turing. Mathematically, ordinal logics present what are probably the most fundamental problems in proof theory. Introduced by Turing [2] in 1939 with the purpose of avoiding as far as possible the effects of Godel's incompleteness theorem, ordinal logics are collections of axiom systems indexed by notations for recursive ordinals. Work of Turing [2] and Feferman [1] yielded that with an ordinal logic using only very small ordinals, one could already decide every mathematical question. This result, however, required the use of arbitrary recursive well-orderings of integers for notations for ordinals, including very wild ones, and thus involved the use of axiom systems that one couldn't reach in any constructive manner. It was also known from Kreisel, Schoenfield, and Wang [1] that if arbitrary well-orderings were allowed, one could prove the consistency of first order arithmetic (and other systems) with ordinals much smaller than ϵ_0.

Considerations including these show the desirability of a natural, or canonical, definition of the recursive ordinals. Cantor provided us with a definition for a subset of these ordinals, and this subset has been expanded by Schütte [4] and others. In all these cases, however, it is clear that the definition does not provide us with all the recursive ordinals. If such a definition could be provided, we might see the Turing-Feferman completeness reversed or turned into a much more valuable completeness result (Gödel believes the latter is likely to happen), see the stumbling block of Myhill [1] overcome, and be able to be sure that the ordinals we assign to formal systems (as described below) make sense.

Gödel once more. In 1958 Gödel [5] made a further important contribution to proof theory, which brought up more clearly the questions suggested by his 1931 and 1933 papers. In this work Gödel provided a new consistency proof for first order number theory with the use of an interpretation of number theory into a free variable equation calculus for certain constructive functions, what are now often called the primitive recursive functionals of finite type. This, thus, was the second interpretation of classical reasoning into constructive reasoning that Gödel had given, and it clearly provided a more revealing explanation for why classical first order reasoning gave correct constructive theorems.

Gödel pointed out that this result added to the evidence (provided by his second incompleteness result and the methods used so far by him, Gentzen and others for consistency proofs for number theory) for the proposition that abstract concepts are needed for the proof of the consistency of number theory (again counter to Hilbert's view). Here abstract concepts were described as thought constructs such as meaningful assertions or proofs, in particular not combinatorial properties of concrete objects. Thus the intuitionistic number theory he had used for his 1933 consistency proof had needed to take meaningful propositions and proofs as basic objects, while to prove transfinite induction up to ϵ_0, Gentzen had had to use the abstract concept of "accessibility." Finally Gödel was obtaining a new consistency proof here through the use of primitive recursive functionals, thus taking functions of functions of functions etc. as his basic objects.

Indeed, Gödel said, after Gentzen's proof of the provability in number theory of all recursions over ordinal numbers less than ϵ_0 (yielding the impossibility of using them alone for a consistency proof), there could no longer practically be any doubt about this. For it is inconceivable that we could give a finitary proof of recursion up to ϵ_0 -- already at ω^ω some of us may be near, or beyond, the limit of what we can justify finitarily.

Thus, Gödel noted, the consistency proofs that had been found had taught us to distinguish two components of the finite attitude, namely the constructive element, according to which one cannot claim to know the existence of an object without producing it or at least providing a method to produce it; and secondly, the specifically finite element, according to which in the last analysis there is a space-time arrangement of the elements one constructs, whose nature is irrelevant except with respect to sameness and distinctness. Thus the consistency results have indicated that the first component must be retained and the second abandoned. Gödel finally suggested that by making these concepts more precise we might be able to obtain proofs of the relationships that actually hold. In particular, we might be able to prove that the use of abstract concepts is actually required for a consistency proof for number theory or for a proof of transfinite induction up to ϵ_0. Thus Gödel made it very clear that he didn't feel that his incompleteness theorems alone had yielded these results in a sufficiently clear way. In this context, Gödel noted, it would be important to distinguish between the concepts of evidence intuitive _for us_ and _idealized_ intuitive evidence, the latter being the evidence which would be intuitive to an idealized finitary mathematician, one who could survey completely finitary processes of arbitrary complexity. _Our_ need for an abstract concept might be due to our inability to understand subject matter that is too complicated combinatorially. By ignoring this, we might be able to obtain an adequate characterization of idealized intuitive evidence. This would not help with Hilbert's program, of course, where we have to use the means at our disposal, but would nevertheless be extremely interesting both mathematically and philosophically. No significant advance seems to have been made toward the solution of these problems to the present time.

Bishop. Taking up Brouwer's rather than Hilbert's program, Bishop [1] and his students have recently succeeded in developing a constructive mathematics much more successfully than Brouwer or the others (such as

Hilbert, Weyl) of his time who had tried. One should note, indeed, that they also have abandoned any restriction of keeping to the specifically finite element (as distinguished by Gödel in 1958). It can be hoped that the elaboration of this work will provide further understanding of the relationship between classical and constructive "truth." Especially at this point when classically trained mathematicians are just beginning to try once more to understand the constructive point of view, the differences between the positions might be expected to stand out more sharply than at any other time, and may provide, if looked at carefully, further information to the philosophers who are trying to understand as much as possible about the nature of and the relationships between these concepts.

Some Comments on the Current Status of the Work in Proof Theory. The active direction of proof theory that is closest to Hilbert's original goal is the one in which people are continuing to look for consistency proofs for various formal systems using means in some sense more elementary than those involved in the formal system in question. Results have been obtained through extensions of Gentzen's use of transfinite induction (e.g. Takeuti [1], [2], [3], [4], Schütte [1], [2], [3], [4], [5], [6], [7], Feferman [1], [2], [3], and Scarpellini [1], [2], [3]).

The methods developed in proof theory and perhaps also its very name have given rise to extensions of its scope beyond what Hilbert conceived of as being meaningful: "proof theory" is now often taken to mean just "the theory of proofs," whatever that might comprise. For instance, out of Gentzen's work came the concept of cut elimination, a concept that clearly has a value of its own independent of Hilbert's program. For one thing, if "cuts" can be eliminated from proofs in a formal system, this means for many formal systems that every provable formula in the system has a proof in a very strong kind of normal form, namely a proof in which all the formulas that occur in the proof are subformulas of the formula being proved. In many formalizations of intuitionistic logic (as Scarpellini has shown, cf. above), moreover, putting a proof into normal form allows one to obtain from a proof of an existential formula $(\exists x)A(x)$ a proof of $A(t)$, for some term t. (A similar statement holds for a disjunction $A \vee B$.) Another important result coming out of the methods of proof theory is that we can attach an ordinal in the natural representation of recursive ordinals to a formal system that measures its strength in a natural way (this is the smallest ordinal one can use to prove the system's consistency). Feferman's work is of special interest here. He was able to clarify the notion of predicativity and to find the ordinal associated with it. It is clear, of course, that for "proof theory" outside Hilbert's program, constructive methods would not be required. We thus have the nonconstructive proof of Prawitz [2] and Takahashi [1] that cut elimination holds for type theory. The recent work of Girard [1], Martin-Löf [1] and Luckhardt [1] is also in this category. Cf. Kreisel [1] and especially [2] for further details.

A final "direction" of proof theory should perhaps be brought out, namely the application of its results and techniques in other parts of logic and mathematics. For instance, much of Bishop's work could be looked upon as applying Gödel's Dialectica interpretation to constructivize classical mathematics (although Bishop worked out his approach without knowledge of Gödel's work). Kreisel, Takeuti, and M. Yasuhara (and others) have also worked on constructivizing classical proofs (recent work in this more logically oriented approach is that of

Takeuti [5] and Yasuhara [1]).

Another direction that has been pursued has involved trying to come to further understanding of the primitive recursive functionals of finite type introduced by Gödel (as well as their extensions). For instance, Tait [2] showed the computation prescribed by any such functional always terminates. Howard [1] then used ordinal assignments to functional terms to show that Tait s result could be obtained by means of first order free variable primitive recursive (i.e. Skolem) arithmetic extended by the descending chain principle for the ordinals less than ϵ_0.

It is clear that work still remains to be done in proof theory, for there are fundamental questions that have not yet been answered. We have proved the consistency of a number of successively more powerful mathematical systems by more and more powerful means. Can we characterize all the possible formal systems we will ever need to use to formalize classical mathematics, and can we further show there is always a constructive proof for such a system? If we can, why can we do this? E.g., is there always a "mirroring" of classical into constructive truth, and if so, what is the precise nature of the mirroring? What is the nature of the relationship between the formal systems and the ordinal corresponding to them through the consistency proofs (and between either of them and the functionals corresponding to them)? Finally, what are the answers to the questions Gödel raised in 1958, questions that are clearly the most fundamental ones if one wishes to consider seriously the epistemological issues involved in the point of view of Hilbert, Brouwer, Gödel and Bishop?

<u>The Contents of this Volume</u>. Most of this work is devoted to presenting aspects of proof theory that have developed out of Gentzen's work. Thus the theme is "cut elimination" and transfinite induction over constructive ordinals. Smullyan's tableau systems will be used for the formalisms and some of the basic logical results as presented in Smullyan [1] will be assumed to be known (essentially only the classical completeness and consistency proofs for propositional and first order logic).

Chapter I presents constructive consistency proofs for first order number theory that are closely related to those of Gentzen [2] and Schütte [5]. The development follows that of Schütte (certain non-constructive aspects of Schütte's work have been remedied in a well-known way).

Chapter II considers proof theoretic and classical topics in pure second order logic and type theory, with an emphasis on the former for the sake of convenience. Here no constructive consistency proof is known for type theory, or even second order logic. In particular, no constructive proof of cut elimination is known (i.e. a constructive consistency proof extending Gentzen's methods). But a classical proof of cut elimination has been found by Prawitz [2] and Takahaski [1] and is given here (in a model-theoretic version, and just for second order logic in order to simplify matters). There are also given two new theorems similar to the cut elimination theorem -- both in structure and in the sense that a constructive proof of either of them would yield a constructive consistency proof for the associated mathematical system -- but with significantly shorter and conceptually simpler classical proofs. Since, however, it seems that it is likely that the constructive proof of the new theorems would not be more elementary than one of ordinary cut elimination, this may be an indication of some further difference between the structure of classical and constructive knowledge.

Other topics considered in this chapter involve the model theory of higher order logics, the relationship between systems formalized with comprehension axioms and those formalized with abstraction terms, and generalized abstract frameworks for higher order logic similar to those considered by Smullyan in [1] (in particular a Henkin completeness proof is given which is simultaneously a completeness proof for first order logic, the usual higher order logics and type theory).

Chapter III completes the proof theoretic treatment of systems equivalent to those considered by Schütte in [5] that was begun in Chapter I.

The first appendix explores further the constructivity of the constructive cut elimination proof for first order logic. It shows that when we eliminate cuts in the usual way from a first order proof, we form a new proof which preserves the "arguments" of the original proof although these arguments may be intertwined and some may be repeated.

The second appendix illustrates translation procedures for going from a proof in a Schütte system to one in a (Smullyan) tableau system and vice versa. Such a procedure is presented only for first order systems since the modifications for higher order systems are easily made.

The third appendix describes briefly the applications of extensions systems since the modifications for higher order systems are easily made.

The third appendix describes briefly the applications of extensions of Gentzen's methods that Scarpellini has made to intuitionistic systems.

To my friend and teacher, Raymond Smullyan, for listening to my thoughts as I learned about proof theory, and to Kurt Gödel, for his many helpful comments on this introduction, I would like to express my deep gratitude.

This work will use many of the concepts developed by Smullyan in [1]. A constructive system of ordinals will also be assumed such as developed by Schütte [1], [2], [3]. We will now review briefly the main concepts developed by those authors that we will be using repeatedly.

Our basic proof-theoretic structure will be a <u>tree</u>. We shall think of trees as having the origin at the top and the successor(s) of each point x below x, in order from left to right and with a line segment from x to them. For any point x in a tree, there will be a unique path P_x (called a branch, if x is the end point of P_x) through the tree from the origin to x. We shall have occasion to speak of adding new points as successors of an end point x of a given tree $\mathcal{T}$, or of extending a branch P_x in $\mathcal{T}$.

A tree is _finitely generated_ if each point has only finitely many successors. A tree is finite or infinite depending on whether it has finitely or infinitely many points.

König's lemma is to the effect that any infinite finitely generated tree has an infinite branch. It is not constructively valid.

Trees constructed according to certain rules will be called tableaux. Some rules will appear in almost all the formal systems we will be concerned with and may be described with a unifying notation. Thus let us briefly consider common features of such systems and rules used to form tableaux in them.

We will deal with _formulas_ that are strings of symbols built up from primitive symbols that usually include variables (of various kinds), parameters, (again of various kinds) and the logical symbols $\neg$, $\wedge$, $\vee$, $\supset$, $\forall$, $\exists$. If we let A^y_x stand for the result of substituting the symbol x for the symbol y in the string A, and assume that we know what certain strings called _atomic formulas_ are, the definition for formulas usually includes the following:

a) Any atomic formula is a formula.

b) If A and B are formulas, $\neg A$, $(A \wedge B)$, $(A \vee B)$, and $(A \supset B)$ · are formulas.

c) If A is a formula and b is a parameter of the same kind as x, $(\forall x)A_x^b$ and $(\exists x)A_x^b$ are formulas.

"Open" and "closed" formulas are defined as usual.

In fact, we will most often be dealing with objects called "signed formulas." Thus we will usually have at hand two <u>signs</u> T and F. We will use Υ to denote an arbitrary sign, and if Υ denotes T (F), then $\overline{\Upsilon}$ will denote F (T, respectively). A <u>signed formula</u> will be a pair ΥA, where A is a formula, A being called the <u>body</u> of the signed formula, and Υ its <u>prefix</u>. If X is a signed formula ΥA, then $\overline{X}$ denotes $\overline{\Upsilon}A$ and $\overline{X}$ is called the <u>conjugate</u> of X. A signed formula is called <u>atomic</u> if its body is atomic. The <u>degree</u> of a signed or unsigned formula is the number of "logical symbols" $\neg$, $\wedge$, $\vee$, $\supset$, $\forall$, $\exists$ it contains that are not within atomic subformulas. In the future the word formula will refer to a closed signed formula unless mention is made to the contrary. (It is only such formulas that will occur as points in tableaux.) We will usually use the letters A, B, C, D to denote unsigned formulas, U, V, W, X, Y to denote signed formulas.

We use the letter "α" to stand for any formula of one of the forms, $T \neg A$, $F \neg A$, $T(A \wedge B)$, $F(A \vee B)$, $F(A \supset B)$, and for each α we define formulas α_1 and α_2 according to the following table:

α	α_1	α_2
T $\neg$ A	F A	F A
F $\neg$ A	T A	T A
T(A$\wedge$B)	T A	T B
F(A$\vee$B)	F A	F B
F(A$\supset$B)	T A	F B

(We will find that in any "interpretation", α will be true iff α_1 and α_2 are both true.) We shall sometimes refer to α formulas as being conjunctive formulas.

We use "β" to stand for any formula of one of the forms F $\neg$ A, T $\neg$ A, F(A$\wedge$B), T(A$\vee$B), T(A$\supset$B), and for each β we define formulas β_1 and β_2 according to the table:

β	β_1	β_2
F $\neg$ A	T A	T A
T $\neg$ A	F A	F A
F(A$\wedge$B)	F A	F B
T(A$\vee$B)	T A	T B
T(A$\supset$B)	F A	T B

(In an "interpretation" β will be true iff at least one of β_1, β_2 is true.) We refer to β formulas as being disjunctive formulas.

We use "γ" to denote any formula of the form T($\forall$x)A(x) or F($\exists$x)A(x) and for any term t,[2] denote by $\gamma(t)$, T A(t) or F A(t) respectively. (In a system with variables of different kinds, the term t must, of course, be of the same kind as the variable x.) Such formulas are called universal formulas.

---

[2] Terms occur in atomic formulas in the same places variables may occur and are defined in different ways in different systems.

We use "δ" to denote any formula of the form $T(\exists x)A(x)$ or $F(\forall x(A(x)$, and for any term t denote by $\delta(t)$, $T\ A(t)$ or $F\ A(t)$ respectively. Such formulas are called <u>existential</u> formulas.

We note the following properties of conjugation:

J_0. (a) $\overline{X}$ is distinct from X.

(b) $\overline{\overline{X}}$ is X.

J_1. (a) The conjugate of any α is a β.

(b) The conjugate of any β is an α.

(c) The conjugate of any γ is a δ.

(d) The conjugate of any δ is a γ.

J_2. (a) $(\overline{\alpha})_1 = \overline{\alpha_1}$; $(\overline{\alpha})_2 = \overline{\alpha_2}$;

(b) $(\overline{\beta})_1 = \overline{\beta_1}$; $(\overline{\beta})_2 = \overline{\beta_2}$;

(c) $\overline{\gamma}(t) = \overline{\gamma(t)}$; $\overline{\delta}(t) = \overline{\delta(t)}$.

Now we consider how this unifying notation is used in describing a tableau system. We describe a tableau system by giving rules for prescribing an origin of a tree and rules for extending branches in a given tree. Thus a tableau for a finite set of formulas S is a tree constructed according to certain rules which <u>usually</u> include:

(A) We may take the branch

$$\begin{array}{c} s_1 \\ \cdot \\ \cdot \\ \cdot \\ s_n \end{array}$$

where $s_i \in S$, $i = 1,\dots,n$, as the tableau. We call this the <u>origin branch</u>.

(B) If some α occurs on the path P_y, we may adjoin either α_1 or α_2 as the sole successor of y.

(C) If some β occurs on the path P_y, we may adjoin β_1 as the first ("left") successor of y and β_2 as the second ("right") successor of y.

(D) If some γ occurs on the path P_y, we may adjoin $\gamma(t)$ as the sole successor of y, where t is any term (of the appropriate kind).

(E) If some δ occurs on the path P_y, we may adjoin $\delta(a)$ as the sole successor of y, where a is any parameter that occurs in no formula of P_y.

(F) We may adjoin X and $\overline{X}$ as the first and second successors to y in P_y.

(G) If X is an element of P_y we may adjoin X as the sole successor of y.

These rules are called respectively the assumption, α, β, γ, δ, cut, and repetition rules. The last rule is only included in tableau systems to make proofs of certain metatheorems somewhat simpler to state. It will always be the case that a tableau using this rule can be transformed to one not using the rule but having all the important properties of the first.

For a cut

$$\begin{array}{ccc} & X & \\ \swarrow & & \searrow \\ \mathfrak{T}B & & \overline{\mathfrak{T}}B \end{array}$$

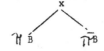

each of B, $\mathfrak{T}B$, and $\overline{\mathfrak{T}}B$ will be called the cut formula of the cut. We say a cut is of degree d if d is the degree of its cut formula. If $\mathcal{T}$ is a tableau, all of whose cuts are of degree less than some integer N, we say $\mathcal{T}$ is of degree 0 if it has no cuts and is of degree d if d is the least upper bound on the degrees of the cuts in it. A tree with a degree, i.e., one for which there is a finite upper bound on the degrees of its cut formulas, is called a _cut-bounded tree_. If X is used as the premise of a rule other than the cut rule in an application of the rule in which the formula Y is obtained, X is the _parent_ of Y and Y is a _direct descendent_ of X in the tableau. Y is a _descendent_ of X if Y is a direct descendent of X or Y is a direct descendent of a descendent of X. X is an _ancestor_ of Y if X is the parent of Y or X is the parent of an ancestor of Y.

Certain finite sets of formulas are distinguished as _closure sets_. (In the so-called pure systems of logic, the closure sets are the sets $\{X, \overline{X}\}$, for X an atomic formula.) A branch of a tableau is _closed_ if it contains all the elements of some closure set; otherwise a branch is called _open_. A tableau is called closed if all its branches are closed. We limit our considerations of closed tableaux to those with finite branches.

We call a set S _inconsistent_ if there is a closed tableau for S. We define a formula X (unsigned formula A) to be _provable_ if the unit set $\{\overline{X}\}$ ($\{F\ A\}$, respectively) is inconsistent, and by a proof of X (A) we mean a closed tableau for $\overline{X}$ (F A).

We say a tableau system is _syntactically consistent_ if there is no (unsigned) formula A such that both A and $\neg A$ are provable.

In our constructive proofs we will frequently want to assign "constructive ordinals" to proofs to allow ourselves to use transfinite induction. Let us now mention what we wish to assume for this purpose.

We assume one can define a binary relation $\otimes$ on the integers for which one can prove constructively:

(a) $\otimes$ is a decidable relation;

(b) $\otimes$ is an order relation, (i.e.,

$a \otimes b$ and $b \otimes c$ implies $a \otimes c$;

for no a does $a \otimes a$ hold;

if $a \neq b$, then $a \otimes b$ or $b \otimes a$);

(c) for any decidable property $P(x)$ of the integers, $P(n)$ holds for all integers n, if, whenever $P(a)$ holds for all integers $a \otimes b$, then $P(b)$ also holds;

(d) there exists no infinite sequence $\sigma_1, \ldots \sigma_n, \ldots$, such that $\sigma_1 \otimes \ldots \otimes \sigma_n \otimes, \ldots$ (this property is implied by property c).

We speak of the natural numbers as <u>ordinals</u> when we have the ordering $\otimes$ in mind. We assume we can find for each ordinal a an ordinal a^* such that there is no ordinal b with $a \otimes b \otimes a^*$. We divide the ordinals into three kinds:

1) the ordinal 0;

2) so-called successor ordinals β for which there is an ordinal $\alpha \otimes \beta$ such that $\beta = \alpha^*$;

3) all other ordinals, which we call <u>limit</u> <u>ordinals</u>.

We assume that we can <u>decide</u> for any ordinal which of these kinds it is. We assume we can exhibit a certain ordinal ω that is a limit ordinal and which is such that for any other limit ordinal γ, $\omega \otimes \gamma$.

We assume further that we can define constructively certain decidable functions and relations on the natural numbers having properties familiar to us from classical ordinal number theory. Thus we have the following (these statements are numbered for later reference):

1. A function $\oplus$ such that:

 a. $a \oplus 0 = a$, $0 \oplus a = a$,

 b. $a \oplus b^* = (a \oplus b)^*$,

 c. $a \otimes b$ implies $a \oplus c \circleddash b \oplus c$,[3]

 d. $b \otimes c$ implies $a \oplus b \circleddash a \oplus c$,

 e. $(a \oplus b) \oplus c = a \oplus (b \oplus c)$.

2. A function E such that:

 a. $E(0) = 1$,

 b. $E(a^*) = E(a) \oplus E(a)$,

 c. $a \otimes b$ implies $E(a) \otimes E(b)$,

 d. $a \circleddash E(a)$.

3. A relation "a is an ϵ-number" and a function ϵ such that:

 a. if a is an ϵ-number, $a \neq \omega$, and $E(a) = a$,

 b. for all integers a, $\epsilon(a)$ is an ordinal number,

 c. for every ϵ-number a, there is an ordinal b such that $\epsilon(b) = a$,

 d. $a \otimes b$ implies $\epsilon(a) \otimes \epsilon(b)$,

 e. $a \otimes \epsilon(a)$.

We sometimes write $\epsilon(a)$ as ϵ_a.

4. A relation "a is a critical ϵ-number" and a function $\mathcal{X}$ such that:

 a. if a is a critical ϵ-number, $\epsilon(a) = a$,

 b. for all integers a, $\mathcal{X}(a)$ is a critical ϵ-number,

 c. for every critical ϵ-number a, there is an ordinal b such that $\mathcal{X}(a) = b$,

---

[3] Here $a \circleddash b$ means $a \otimes b$ or $a = b$.

d. $a \mathbin{\textcircled{<}} b$ implies $\mathcal{K}(a) \mathbin{\textcircled{<}} \mathcal{K}(b)$,

e. $a \mathbin{\textcircled{\le}} \mathcal{K}(a)$.

5. A function $\odot$ such that:

a. $a \odot 0 = 0,\ 0 \odot a = 0$,

b. $a \odot b^* = (a \odot b) \oplus a$,

c. $a \mathbin{\textcircled{<}} b$ implies $a \odot c \mathbin{\textcircled{\le}} b \odot c$,

d. $a \neq 0$ and $b \mathbin{\textcircled{<}} c$ implies $a \odot b \mathbin{\textcircled{<}} a \odot c$,

e. $a \odot (b \oplus c) = (a \odot b) \oplus (a \odot c)$,

f. $(a \odot b) \odot c = a \odot (b \odot c)$,

g. $E(a \oplus b) = E(a) \odot E(b)$.

6. A function a^b such that:

a. $a^0 = 1$,

b. $a^{b^*} = a^b \odot a$,

c. $E(a) = 2^a$,

d. $a \mathbin{\textcircled{<}} b$ implies $a^c \mathbin{\textcircled{\le}} b^c$,

e. $1 \mathbin{\textcircled{<}} a$ and $b \mathbin{\textcircled{<}} c$ implies $a^b \mathbin{\textcircled{<}} a^c$,

f. $1^a = 1$,

g. $a^{b \oplus c} = a^b \odot a^c$,

h. $a^{b \odot c} = (a^b)^c$,

i. $1 \mathbin{\textcircled{<}} a$ implies $b \mathbin{\textcircled{\le}} a^b$.

7. A function $a_n(b)$ such that:

a. $a_0(b) = b$,

b. $a_{n+1}(b) = a^{a_n(b)}$,

c. $a \mathbin{\textcircled{<}} b$ implies $a_n(c) \mathbin{\textcircled{\le}} b_n(c)$,

d. $1 \mathbin{\textcircled{<}} a$ and $b \mathbin{\textcircled{<}} c$ implies $a_n(b) \mathbin{\textcircled{<}} a_n(c)$,

e. $1 \mathbin{\textcircled{<}} a$ and $m < n$ implies $a_m(b) \mathbin{\textcircled{\le}} a_n(b)$,

f. $1 \mathbin{\textcircled{<}} a$ implies $b \mathbin{\textcircled{\le}} a_n(b)$,

g. $a_m(a_n(b)) = a_{m+n}(b)$.

8. A function ψ such that:

a. $a \textcircled{<} b+2^{\psi(a,b)*}$,

b. $c \neq 0$ and $a \textcircled{<} b \oplus 2^c$ implies $\psi(a,b) \textcircled{<} c$.

9. A function μ such that:

a. $\mu(0) = 0$,

b. $\mu(z') = \mu(z)*$.

10. A function ν such that:

a. $\nu(0) = 0$,

b. $\nu(z*) = \nu(z')$,

c. $\nu(z_\lambda) = 0$ for every limit number z_λ.

11. A function $L(a)$ such that:

a. $L(a)$ is ω if $a \textcircled{<} \epsilon_0$,

b. if $\epsilon_0 \textcircled{\leq} a$, $L(a)$ is an ϵ-number, $L(a) \textcircled{\leq} a$ and there is no ϵ-number b such that $L(a) \textcircled{<} b \textcircled{\leq} a$ (i.e., classically here $L(a)$ is the largest ϵ-number $\textcircled{\leq} a$).

12. A function $N(a,b)$ such that:

a. $1 \textcircled{<} b$ and $N(a,b) = n$ implies $a \textcircled{<} b_n(L(a)*)$.

13. A function $\#$ called the <u>natural</u> <u>sum</u> such that:

a. $a \# 0 = a$, $0 \# a = a$,

b. if $a = \omega\mathcal{H}^0 \odot \mu_0 \oplus \dots \oplus \omega\mathcal{H}^n \odot \mu_n$,

$b = \omega\mathcal{H}^0 \odot \nu_0 \oplus \dots \oplus \omega\mathcal{H}^n \odot \nu_n$,

with $\mathcal{H}_n \textcircled{<} \dots \textcircled{<} \mathcal{H}_0$, $\mu_i \textcircled{<} \omega$, $\nu_i \textcircled{<} \omega$ for all $i = 0, \dots, n$, then

$a \# b = \omega\mathcal{H}^0 \odot (\mu_0 \oplus \nu_0) \oplus \dots \oplus \omega\mathcal{H}^n \odot (\mu_n \oplus \nu_n)$.

c. $a \# b = b \# a$,

d. $(a \# b) \# c = a \# (b \# c)$,

e. $a \textcircled{<} b$ implies $a \# c \textcircled{<} b \# c$,

f. $a \neq 0$ implies $a \# a \# \ldots \# a \textcircled{<} a \odot \omega$,

g. $n > 0$ and $a \textcircled{<} b$ implies $\omega_n(a) \# \omega_n(a) \textcircled{<} \omega_n(b)$.

14. A function ρ such that:

a. $\rho(a,b) \textcircled{<} \rho(c,b)$ for $a \textcircled{<} c$,

b. $\rho(a,b) \textcircled{<} \rho(a,c)$ for $a \textcircled{<} c$,

c. $\rho(\epsilon(\rho(a,b) \oplus 1),\ c) \textcircled{<} \rho(d,b)$ for

$a \textcircled{<} d$ and $c \textcircled{<} b$.

We are also able to prove constructively:

15. If ϵ is an ϵ-number,

a. $a \textcircled{<} \epsilon$ and $b \textcircled{<} \epsilon$ implies

$a \oplus b \textcircled{<} \epsilon$,

$a \odot b \textcircled{<} \epsilon$,

$a^b \textcircled{<} \epsilon$,

$a_n(b) \textcircled{<} \epsilon$,

$a \# b \textcircled{<} \epsilon$.

b. $1 \textcircled{<} a \textcircled{<} \epsilon$ implies $a \oplus \epsilon = a \odot \epsilon = a^{\epsilon} = a_n(\epsilon) = \epsilon$.

We will often associate ordinals with formulas in finite-branched tableaux and also with the tableaux themselves. We will usually do this as follows. Let X be an occurrence of a formula in a finite-branched tableau. If X is an end point of a branch, and μ is any ordinal, X is of rank $\leq \mu$. Let X be a point with successors $Y_1, \ldots Y_n, \ldots$. If for each n there is a $\mu_n < \mu$ such that the rank of X_n is $\leq \mu_n$, then the rank of X is $\leq \mu$. Let $\mathcal{T}$ be a finite-branched tableau with origin branch

$$\begin{array}{c} s_1 \\ \cdot \\ \cdot \\ \cdot \\ s_n \end{array}$$

Then we say the <u>rank</u> of $\mathcal{T}$ is $\leq \mu$ if the rank of s_n is $\leq \mu$.

We say a formula (tableau) is of rank $< \nu$ if there is some ordinal $\mu < \nu$ such that the formula (tableau) is of rank $\leq \mu$.

There may be a formula X in some tableau $\mathcal{T}$ for which there is no ordinal μ such that X is of rank $\leq \mu$ in $\mathcal{T}$. We will use the following two lemmas about ranks.

<u>Rank Lemma 1</u>. Let σ be an arbitrary ordinal $\textcircled{>} 1$. Let X_i be the set of successors of X in a finite-branched tree $\mathcal{T}$, and let the rank of each X_i be $\leq \sigma^{\alpha_i}$. Then if $\alpha_i \textcircled{<} \alpha$ for all i, the rank of X in $\mathcal{T}$ is $\leq \sigma^{\alpha}$.

Proof: We need only show $\sigma^{\alpha_i} \textcircled{<} \sigma^{\alpha}$ for all i. This is true by 6e.

<u>Corollary</u>. Let σ be an arbitrary ordinal. Let $\mathcal{T}$ be a tree in which X has successors Y_i with ranks $\leq \alpha_i$, and let $\alpha_i \textcircled{<} \alpha$ for all i. If $\mathcal{T}$ is transformed into a tree $\mathcal{T}'$ in which X has the same successors Y_i, but in which the rank of each Y_i is $\leq \sigma^{\alpha_i}$, the rank of X in the new tree $\mathcal{T}'$ is $\leq \sigma^{\alpha}$.

<u>Rank Lemma 2</u>. Let $\mathcal{T}$ be a tree of rank $\leq \alpha$. Let $\mathcal{T}'$ be a tree that contains $\mathcal{T}$ as a subtree in such a way that the only points of $\mathcal{T}'$ not in $\mathcal{T}$ occur below end points of $\mathcal{T}$. Then if the ranks of the end points of $\mathcal{T}$ in $\mathcal{T}'$ are all $\leq \beta$, the rank of $\mathcal{T}'$ is $\leq \beta \oplus \alpha$.

Proof. We use transfinite induction over ordinals $\gamma \textcircled{<} \alpha$, showing the proposition for any point X of rank $\leq \gamma$ in $\mathcal{T}$. If γ is 0, X must be an end-point of $\mathcal{T}$ and by assumption the rank of X in $\mathcal{T}'$ is $\leq \beta = \beta \oplus \gamma$.

Let γ be greater than 0 and assume the lemma true for ordinals smaller than γ. If X is an end point in $\mathcal{T}$, the rank of X is $\leq \beta$ in $\mathcal{T}'$ by assumption and thus also $\leq \beta \oplus \gamma$. Now consider when X has successors X_i with ranks $\alpha_i \circledless \gamma$ in $\mathcal{T}$. By the induction hypothesis the ranks of the X_i are $\leq \beta \oplus \alpha_i$ in $\mathcal{T}'$. But $\beta \oplus \alpha_i \circledless \beta \oplus \gamma$ when $\alpha_i \circledless \gamma$ by statement 1d. So the rank of X in $\mathcal{T}'$ is $\leq \beta \oplus \gamma$.

CHAPTER I - FIRST ORDER NUMBER THEORY

In this chapter we consider two systems for first order number theory. Here formulas are built up as in first order logic, but starting also with atomic formulas consisting of decidable number-theoretic predicates applied to terms involving decidable number theoretic functions. Closure conditions are added which are based on the new "meanings" of the atomic formulas. In the first system, Ψ,[1] we add a rule of complete induction, while in the second system, Ω, we do away with individual parameters completely and add an infinitary γ rule, where from γ we branch to $\gamma(n)$, for each numeral n.

Various constructive proofs of the consistency of a system like Ψ have been given, for instance, those of Gentzen [1], [2], Ackermann [1], Takeuti [1], [2], and Gödel [5]. In §2 we will present a proof based on that of Gentzen [2]. In §4 we will present a consistency proof of Ω that is similar to the Hauptsatz for first order logic (for other consistency proofs for systems with infinitary rules, see e.g., Lorenzen [1], Schütte [1], [2], [3], Stenius [1], Ackermann [2], Tait [2]).

Let us now proceed with the defining of the two systems.

§1. The Finitary System Ψ

We define the numerals to be those symbol strings that can be formed with the symbol 0 and the apostrophe according to the following rules:

---

1 For convenience of comparison we use names here that are the same as those for similar systems described by Schütte.

a) The number symbol 0 is a numeral.

b) If n is a numeral, so is n'.

The numerals represent the natural numbers 0, 1, 2, ..., where the apostrophe is used to denote the successor function.

An n-place number-theoretic <u>function</u> is a correspondence between every n-tuple of numerals and a numeral. If such a function is denoted by f, then $f(z_1,\ldots,z_n)$ denotes the numeral that corresponds to the n-tuple of numerals $\langle z_1,\ldots,z_n\rangle$ under the function in question. The numeral denoted by $f(z_1,\ldots,z_n)$ will be called the <u>numerical value</u> of the function f at the argument $\langle z_1,\ldots,z_n\rangle$.

An n-place number-theoretic <u>predicate</u> is a correspondence between every n-tuple of numerals and a truth value (t or f, or truth and falsity). If such a predicate is denoted by p, then $pz_1,\ldots,z_n$ denotes the truth value of the predicate at the argument $\langle z_1,\ldots,z_n\rangle$.

We will consider only primitive recursive number-theoretic functions and predicates which we will call <u>evaluable</u> since for each such function or predicate there is a general procedure by which one can calculate the numerical or truth value for every argument. Examples of evaluable functions and predicates are the successor function, the sum function, the product function, the power function, the equality relation (or predicate), the usual order relation on the numerals, the functions and relations on ordinals described in the introduction.

The <u>primitive symbols</u> of Ψ are:

a) The symbol 0, denumerably many parameters, denumerably many variables,

b) Function symbols, predicate symbols, predicate variables,

c) The propositional operators $\neg$, $\wedge$, $\vee$, and $\supset$, and the quantifiers $\forall$ and $\exists$.

d) The comma and right and left parentheses.

Every function symbol, predicate symbol, and predicate variable has a natural number $n \geq 1$ associated with it, the number of its "places." All function and predicate symbols of the system Ψ are symbols for evaluable functions and predicates. Most of the discussion will require only that the successor function and equality relation are represented in the system. Later parts of the discussion will require all the functions and predicates mentioned in the set of examples above, in addition to a few others.

The terms of the system Ψ are those strings of symbols that can be built up through the application of the following two rules:

a) The number symbol 0 and every parameter is a term,

b) If f is an n-place function symbol and $t_1, \ldots, t_n$ are terms, then $f(t_1,\ldots,t_n)$ is also a term.

Terms using familiar functions (e.g., the successor, sum) will be abbreviated in the usual way (i.e., t' for $'(t)$, etc.). The same will be true for familiar predicates (e.g., we will write $a = b$ instead of $= a,b$).

We define an atomic formula to be a symbol string of the form

$$pt_1,\ldots,t_n$$

where p is an n-place predicate symbol and $t_1, \ldots, t_n$ are terms.

We define the formulas of the system Ψ to be those strings of symbols built up from atomic formulas in the usual way, i.e.,

a) Every atomic formula is a formula.

b) If A is a formula, so is $\neg A$.

c) If A and B are formulas, so are $(A \wedge B)$, $(A \vee B)$ and $(A \supset B)$.

d) If A is a formula, then $(\forall x)A$ and $(\exists x)A$ are also formulas.

As non-formal expressions we use:

$a, a_1, \ldots, b, \ldots, c, \ldots$	for parameters,
$x, x_1, \ldots, y, y_1, \ldots$	for variables,
$z, z_1, \ldots$	for numerals,
$s, s_1, \ldots, t, t_1, \ldots$	for terms,
$A, B, C, \ldots, P, Q, \ldots$	for formulas.

We define subformulas, signed formulas, the degree of a formula, a closed formula, etc. as usual (see the introduction). We deal only with closed formulas.

We call a term <u>numerical</u> when it contains no parameters. Every numerical term can be assigned a specific numerical value in the following way:

a) The numeral z has the numerical value z.

b) If the numerical terms $t_1, \ldots, t_n$ have the numerical values $z_1, \ldots, z_n$ respectively, and f is the symbol for an n-place function that associates the numerical value z with the n-tuple $\langle z_1,\ldots,z_n\rangle$, then the numerical value of the term $f(t_1,\ldots,t_n)$ is z.

We say that two (not necessarily numerical) terms or formulas are <u>equivalent</u> if the first can be transformed into the second by replacing numerical terms within the first term or formula with terms having the same numerical value. We say that two terms t_1, t_2 or two formulas A_1, A_2 are <u>equivalent relative to $s = t$</u> (where s and t are arbitrary terms) if t_1 or A_1 can be transformed into a term or formula equivalent to t_2 or A_2 by replacing some occurrences of t by s and some occurrences of s by t (in t_1 or A_1).

A formula (signed or unsigned) is called numerical when it contains no quantifiers and no parameters and no predicate variables. Such formulas can be given a specific truth value in a natural way. We will give the following definition for signed numerical formulas, using the unifying notation due to Smullyan (see the introduction):

a) If the numerical terms $t_1, \ldots, t_n$ have numerical values $z_1, \ldots, z_n$ and p is an n-place predicate symbol, then the numerical atomic formula $T\ pt_1,\ldots,t_n$ has the truth value that corresponds to the n-tuple $<z_1,\ldots,z_n>$ according to the predicate represented by p and $F\ pt_1,\ldots,t_n$ has the opposite truth value.

b) An α formula is true (has the value t) if both α_1 and α_2 are true (have the value t) and otherwise is false (has the value f).

c) A β formula is true (has the value t) if at least one of the two formulas β_1 and β_2 is true (has the value t), and is false (has the value f) otherwise.

Note that this is an inductive definition using the degree of a signed formula, for α_1 and α_2 always have a lower degree than α and β_1 and β_2 always have a lower degree than β. Because all function symbols and predicate symbols of the system belong to evaluable functions and predicates, the numerical values of all numerical terms and the truth values of all numerical formulas can be successively determined by the use of the rules above. Now an unsigned numerical formula P is defined to have the truth value t (f) if $T\ P$ has the truth value t (f). It can be proved inductively using the properties of conjugation that P has the value t (f) iff $F\ P$ has the value f (t).

We note that the following proposition is easy to verify:

a) The relations "is equivalent to" and "is equivalent relative to $s = t$" are equivalence relations, i.e., they are reflexive, symmetric, and transitive;

b) If two terms s, t or formulas P, Q are equivalent and numerical, they have the same numerical or truth values.

A numerical specialization of a formula with no quantifiers and no predicate variables is a formula which results when every parameter is replaced by a numeral, occurrences of the same parameter being replaced by the same numeral. By what was shown above, every such numerical specialization has a specific truth value. Let S be a set of formulas in which no predicate variables or quantifiers occur. A numerical specialization of S is a set that results from S when every parameter occurring in a formula of S is replaced by a numeral, occurrences of the same parameter being replaced by the same numeral. By what was shown above, every formula in a numerical specialization of S has a specific truth value. A set S of such formulas is called arithmetically unsatisfiable if every numerical specialization of it contains a false formula. A formula P is called arithmetically true if every numerical specialization of it is true.

We can now describe the system Ψ of which we will speak.

The rules of Ψ are the α, β, γ, δ, assumption, repetition, and cut rules (see the introduction), and, in addition, the rule of complete induction: if for some unsigned formula A and term t, $F\ A(0) \supset A(t)$ occurs on the branch P_y, then we may adjoin

F $A(a) \supset A(a')$ as the sole successor of Y, for any parameter a not occurring in the branch P_y. We may visualize these as follows:

<u>Rules of Ψ</u>

<u>α</u>

$$\frac{\alpha}{\alpha_1} \qquad \frac{\alpha}{\alpha_2}$$

<u>β</u>

$$\frac{\beta}{\beta_1 \quad \beta_2}$$

<u>γ</u>

$$\frac{\gamma}{\gamma(t)} \quad t \text{ any term}$$

<u>δ</u>

$$\frac{\delta}{\delta(a)} \quad , \quad a \text{ new}$$

<u>Complete Induction</u>

$$\frac{F\ A(0) \supset A(t)}{F\ A(a) \supset A(a')} \quad t \text{ any term, } a \text{ new}$$

<u>Cut</u>

$$\frac{X}{Y \quad \bar{Y}}$$

<u>Assumption</u>

s_1

.

. $s_i \in S$

. $i = 1, \ldots, n$

s_n

<u>Repetition</u>

$$\frac{X}{X}$$

The closure sets of Ψ are the following:

a) the arithmetically unsatisfiable finite sets of atomic formulas;

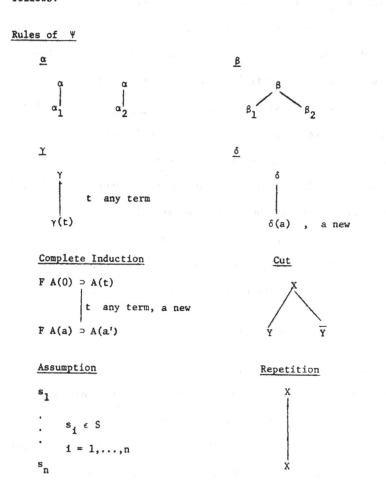

b) the sets $\{A, \overline{A'}\}$ where A and A' are equivalent atomic formulas;

c) the sets $\{T\ s = t, A, \overline{A'}\}$ where A and A' are atomic formulas that are equivalent relative to $s = t$.[2]

We consider ourselves given a proof in Ψ only when we have given (1) either a specific finite tableau or a constructive procedure to obtain such a tableau, and (2) a constructive proof that each branch of the tableau is closed, i.e., that some finite set of atomic formulas on each branch is a closure set.

Notice that by following the constructivists Gentzen and Schütte in letting every finite arithmetically unsatisfiable set of atomic formulas be a closure set, we are now not necessarily able to decide whether a given finite tableau actually is a proof. For instance, at the present we do not know whether the Fermat formula

$$T \quad p > 2 \wedge a^p + b^p = c^p$$

is arithmetically unsatisfiable, so that if a tableau branch included it, we would not be able to decide whether or not the formula could be used to close the branch. It will be clear in the sequel that this absence of a formal system in the strict sense does not prevent one from showing that no false finitistic formula can be proved in the system. (The closure sets, of course, include the negations of the Peano axioms and of recursion equations for the evaluable functions of Ψ, so one can at least prove in Ψ all the theorems that can be proved in the usual formal systems for first order arithmetic. As

---

[2] We note that b) and c) are only necessary because we allow predicate variables in our system. This will be convenient in our incompleteness proof later.

Jesse Wright noted, the classical logician can actually use the results of Gödel and Church about formal systems to show that Ψ is not a formal system. This is done as follows.

We first notice that if A were a formula without quantifiers and predicate variables and the set of closure conditions were recursive, we could decide whether or not A were arithmetically unsatisfiable -- this can easily be proved by induction. Now consider the formal system of number theory described in Kleene [1], say. We know there is a quantifier-free proof predicate $B(a,b)$ containing only the functions $'$, $+$, $\cdot$ and the predicate $=$, and meaning "a is the Gödel number of a proof of the formula with Gödel number b." If the closure sets of Ψ were recursive we could decide for every numeral z whether all numerical specializations of

$$\neg \beta(a, z)$$

were true, i.e., we could decide for any formula whether or not it had a proof. But we know that this is impossible.

The following is an example of a proof of $(\forall y)(\exists x)(y \cdot y' = x + x)$ in Ψ:

1 $F(\forall y)(\exists x)(y \cdot y' = x + x)$

cut

2 $T(\exists x)(0 \cdot 0' = x + x)$ $\qquad$ $F(\exists x)(0 \cdot 0' = x + x)$

δ from 1 $\qquad$ γ

3 $F(\exists x)(c \cdot c' = x + x)$ $\qquad$ $\underline{\underline{F\ 0 \cdot 0' = 0 + 0}}$

cut

4 $T(\exists x)(0 \cdot 0' = x + x) \supset (\exists x)(c \cdot c' = x + x)$

β

5 $T(\exists x)(c \cdot c' = x + x)$

δ

6 $T\ c \cdot c' = e + e$

γ from 3

7 $\underline{\underline{F\ c \cdot c' = e + e}}$

8 $F(\exists x)(0 \cdot 0' = x + x)$

γ

9 $\underline{\underline{F\ 0 \cdot 0' = 0 + 0}}$

10 $F(\exists x)(0 \cdot 0' = x + x) \supset (\exists x)(c \cdot c' = x + x)$

complete induction

11 $F(\exists x)(b \cdot b' = x + x) \supset (\exists x)(b' \cdot b'' = x + x)$

α

12 $T(\exists x)(b \cdot b' = x + x)$

α from 11

13 $F(\exists x)(b' \cdot b'' = x + x)$

δ from 12

14 $T\ b \cdot b' = d + d$

γ from 13

15 $\underline{\underline{F\ b' \cdot b'' = (d + b') + (d + b')}}$

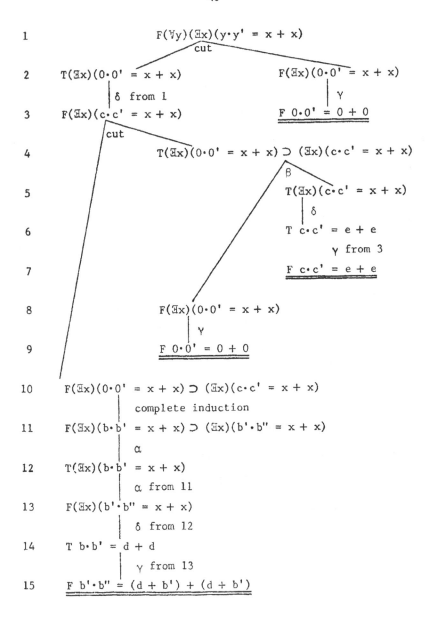

(Here a double underline marks the end of a closed branch. The rules used are indicated and the line number of a premise formula is given when it is not directly above its conclusion.)

We will now prove certain propositions about the system Ψ that we will need.

<u>Proposition 1</u>. If S is any arithmetically unsatisfiable finite set of formulas, S has a closed tableau using only α, β, assumption and repetition rules.

Proof. Let S be a arithmetically unsatisfiable finite set of formulas[3] and define the <u>degree</u> of S to be the sum of the degrees of the formulas in S. We prove the proposition by induction on the degree of S. If S is of degree 0, all the formulas in S are atomic and S is a closure set. Thus if $s_1,\ldots,s_n$ are the elements of S,

$$\begin{array}{c} s_1 \\ \cdot \\ \cdot \\ \cdot \\ s_n \end{array}$$

is a closed tableau for S.

If S is not of degree 0, it must contain some formula X of degree > 0. X must be an α or β (since no quantifiers can occur in any formula of an arithmetically unsatisfiable set). If X is an α, the set $S' = S \cup \{\alpha_1, \alpha_2\} - \{\alpha\}$ is a set of

[3] I.e., a set for which we have a constructive proof that it is arithmetically unsatisfiable.

degree less than S. We can see by considering definitions that S' must be arithmetically unsatisfiable whenever S is. For to know that S is arithmetically unsatisfiable is to know that for any numerical specialization S^* of S, we can find a formula of S^* that is false. If in a certain collection of such numerical specializations we know that α^* is false, we can only know this if we can point out for each specialization which one of α_1^* or α_2^* is false. Now by the induction hypothesis, S' has a closed tableau $\mathcal{T}'$ using only α, β assumption and repetition rules. Then

$$\begin{array}{c}\alpha\\ |\\ \mathcal{T}'\end{array}$$

is a closed tableau for S using only those rules. The case when X is a β formula is similar.

Proposition 2. If S is a set of formulas containing formulas A, $\overline{A'}$, where A and A' are equivalent, S has a closed tableau in which cut and induction rules are not used.

Proof. If A and A' are equivalent, they differ only in numerical terms of the same numerical values. Thus we may say we have a set S containing two formulas $X(t_1,\ldots,t_n)$, $\overline{X(t_1',\ldots,t_n')}$, where t_i and t_i', $i = 1,\ldots,n$, have the same numerical value. We prove by an induction on degree that S has a closed tableau without any uses of the cut and induction rules and in which all the members $s_1,\ldots,s_k$ of S occur as assumptions (this strengthening of the induction statement enables us to forget about which constants δ rules may introduce). If $X(t_1,\ldots,t_n)$ has degree 0, it is atomic, and clearly $\{X(t_1,\ldots,t_n), \overline{X(t_1',\ldots,t_n')}\}$ is a closure set by closure condition b. Thus the tableau

$$\begin{array}{c} s_1 \\ \vdots \\ s_k \end{array}$$

is a closed tableau for the set S that does not use the proscribed rules.

Now let $X(t_1,\ldots,t_n)$ be of degree greater than 0 and assume the proposition true for formulas of lower degree. We have either an α, β pair or a γ, δ pair. Let $X(t_1,\ldots,t_n)$ be an α. Then $\overline{X(t_1',\ldots,t_n')}$ is a β. Let S_i, $i = 1, 2$ be the set $S \cup \{X(t_1,\ldots,t_n)_1, X(t_1,\ldots,t_n)_2, \overline{X(t_1',\ldots,t_n')_i}\}$. S_i contains $X(t_1,\ldots,t_n)_i$ and $\overline{X(t_1',\ldots,t_n')_i}$, and thus has a closed tableau $\mathcal{T}_i$ of the required type by the induction hypothesis. Then

$$\begin{array}{c} s_1 \\ \vdots \\ s_k \\ | \\ X(t_1,\ldots,t_n)_1 \\ | \\ X(t_1,\ldots,t_n)_2 \\ \swarrow \qquad \searrow \\ \overline{X(t_1',\ldots,t_n')_1} \qquad \overline{X(t_1',\ldots,t_n')_2} \\ \mathcal{T}_1 \qquad\qquad\qquad \mathcal{T}_2 \end{array}$$

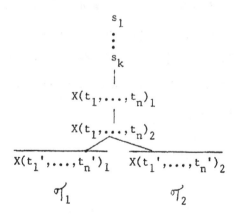

is a closed tableau for S using only the appropriate rules. Now let $X(t_1',\dots,t_n')$ be a γ, so that $\overline{X(t_1',\dots,t_n')}$ is a δ. Let S' be the set $S \cup \{X(t_1,\dots,t_n)(b), \overline{X(t_1',\dots,t_n')(b)}\}$, where b is a parameter and not occurring in S. By the induction hypothesis again, there is a closed tableau $\mathcal{T}'$ of the required type for S'. Hence

$$\begin{array}{c} s_1 \\ \vdots \\ s_k \\ \hline \overline{X(t_1',\dots,t_n')(b)} \\ X(t_1,\dots,t_n)(b) \\ \mathcal{T}' \end{array}$$

is a closed tableau for S not using cut or induction rules.

<u>Proposition</u> 3. If S is a set of formulas containing formulas $T\ s = t$, A, $\overline{A'}$ where A and A' are equivalent relative to $s = t$, S has a closed tableau not using the cut and induction rules.

Proof. Again it is clear that A and A' differ only in terms (and thus have the same logical structures and degrees). We prove the proposition similarly to proposition 2. If A is of degree 0, we see that $\{T\ s = t, A, \overline{A'}\}$ forms a closure set by closure condition c and the tableau

$$\begin{array}{c} T\ s = t \\ | \\ A \\ | \\ \overline{A'} \end{array}$$

is a closed tableau for the set S not using the proscribed rules.

If A is not of degree 0, one of A and $\overline{A'}$ is an α or a γ, the other a β or a δ. In the α-β case, for $i = 1,2$ the set $S_i = \{T\ s = t,\ \alpha_i,\ \beta_i\}$ also satisfies the assumptions of the proposition, so that by the induction hypothesis S_i has a closed tableau $\mathcal{T}_i$ not using cut and induction rules. Thus,

$$\begin{array}{ccc} & s_1 & \\ & \vdots & \\ & s_k & \\ / & & \backslash \\ \beta_1 & & \beta_2 \\ | & & | \\ \alpha_1 & & \alpha_2 \\ | & & | \\ \mathcal{T}_1 & & \mathcal{T}_2 \end{array}$$

is a closed tableau for S not using the proscribed rules. The γ-δ case is similar.

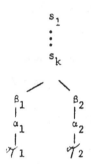

Corollary. If $\mathcal{T}$ is a tableau each of whose branches contains a finite set S of formulas such that either a) S is arithmetically unsatisfiable, or b) S contains formulas A, $\overline{A'}$, where A and A' are equivalent, or c) S contains formulas T s = t, A, $\overline{A'}$, where A and A' are equivalent relative to s = t, then $\mathcal{T}$ can be extended to a closed tableau without the use of cut or induction rules.

Clearly we could add a condition d) to the corollary: S contains a formula X that has a closed tableau; in this case, however the tableau $\mathcal{T}$ might only be able to be extended to a closed tableau if the cut and induction rules are allowed. When we display a tableau in Ψ, we will, sometimes, indicate by putting $\$_a$, $\$_b$, $\$_c$, or $\$_d$ at the end of a branch that the branch satisfies condition a, b, c, or d above (and can thus be extended to a set of closed branches). We call these derived closure conditions a, b, c and d, and say a branch satisfying one of them is preclosed.

Let a rule be given with a finite set of premises A, B, ... and a (finite) set of (open) conclusion branches $\theta_1, \ldots, \theta_n$, where the branch θ_k contains the finite set of formulas $A_1, \ldots, A_k$. We say that such a rule <u>is derivable</u> in Ψ if whenever we have a tableau with a branch containing the premises of the rule, we can, using only finitely many applications of rules of Ψ, extend the branch to a subtree such that there is a 1-1 correspondence between the <u>open</u> branches of the subtree and the branches θ_i of the rule such that each open branch in the subtree contains all the formulas of the branch in the rule corresponding to it.

<u>Proposition 4</u>. The rule $\frac{Y}{X}$, where Y is any formula and X is an arithmetically true signed formula, is a derivable rule in Ψ.

Proof. If X is arithmetically true, $\overline{X}$ is arithmetically unsatisfiable. Thus one can cut with the formulas X and $\overline{X}$ after which the branch containing $\overline{X}$ can be extended to a closed branch by proposition 1. This rule will be called the <u>axiom rule</u>.

<u>Proposition 5</u>. The rule $\frac{\mathcal{T}\ P}{\mathcal{T}\ Q}$, where P and Q are equivalent is a derivable rule of Ψ.

Proof. For one can cut to $\mathcal{T}Q$ and $\overline{\mathcal{T}}Q$, where the branch with $\overline{\mathcal{T}}Q$ can be extended to a closed subtree by proposition 2. This rule will be called the <u>replacement rule</u>.

<u>Proposition 6</u>. The rule $\dfrac{\begin{matrix} T\ s = t \\ \mathcal{T}\ P \end{matrix}}{\mathcal{T}\ Q}$, where s and t are any terms and P and Q are equivalent relative to $s = t$, is a derivable rule in Ψ.

Proof. For one can cut to $\mathcal{T}\ Q$ and $\overline{\mathcal{T}}\ Q$, where the branch with $\overline{\mathcal{T}}\ Q$ can be extended to a closed subtree by proposition 3. This rule will be called the <u>equality rule</u>.

<u>Proposition 7</u>. The rule $\dfrac{Y}{TP}$, where Y is any formula and P is known to have a proof, is a derivable rule in Ψ.

Proof. Obvious. This rule will be called the <u>theorem rule</u>.

<u>Note</u> that only α, β, γ, δ and cut rules are needed to derive the conclusions from the premises in any of the derived rules of propositions 4 - 6.

In displaying our tableaus of Ψ, we will sometimes indicate rules with the following abbreviations:

α	α
β	β
γ	γ
δ	δ
cut	CUT
complete induction	CI
axiom	AX
replacement	REP

equality	EQ
Theorem	TH
Repetition, where an abbreviation is being indicated	AB

We note that since the system Ψ has the α and β rules, and in it a branch with a TP and FP can be extended to a closed branch, the system Ψ can be shown to be propositionally complete by the usual propositional completeness proofs for tableaux (see Smullyan [1]). Because of this, if you can prove a propositionally false formula P, you can prove any formula Q as follows:

there exists a closed tableau for F P by assumption

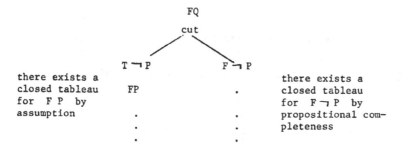

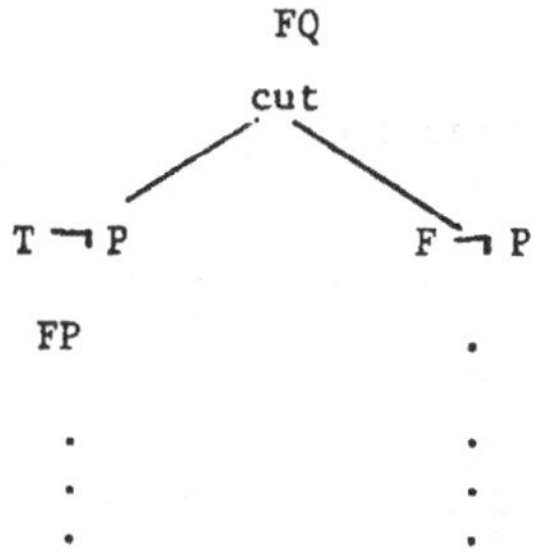

there exists a closed tableau for F ¬ P by propositional completeness

Also, if you could prove both a formula and its negation, by using the same type tableau you could prove any formula Q.

In order to get a better feeling for complete induction in the system, we prove two propositions regarding it.

Proposition 8. For any formula A(a) and any term t,

$$(A(0) \wedge (\forall x)(A(x) \supset A(x'))) \supset A(t)$$

is a provable formula.

Proof. The tableau

$$T(A(0) \land (\forall x)(A(x) \supset A(x'))) \supset A(t)$$

$$\alpha$$

$$T\ A(0) \land (\forall x)(A(x) \supset A(x'))$$

$$\alpha$$

$$F\ A(t)$$

$$\alpha$$

$$T\ A(0)$$

$$\alpha$$

$$T(\forall x)(A(x) \supset A(x'))$$

$$T\ A(0) \supset A(t) \qquad \text{CUT} \qquad F\ A(0) \supset A(t)$$

$$\beta \qquad\qquad\qquad\qquad\qquad CI\ |$$

$$F\ A(0) \quad T\ A(t) \qquad\qquad\qquad F\ A(a) \supset A(a')$$

$$\$_b \qquad\quad \$_b \qquad\qquad\qquad\qquad T\ A(a) \supset A(a')$$

$$\$_b$$

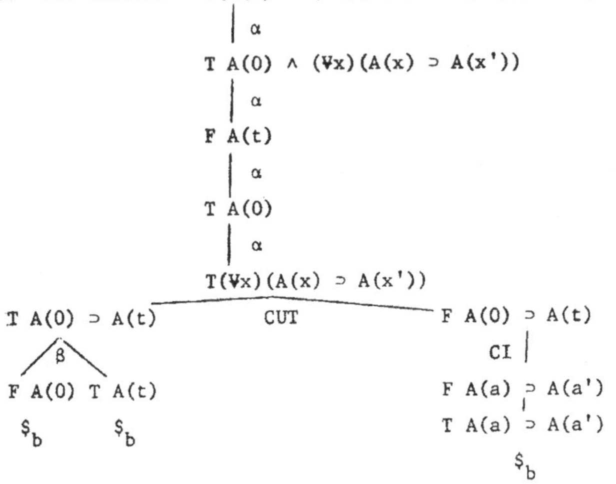

can be extended to a closed tableau.

Proposition 9. For any formula A, sign $\mathcal{T}$, and term t,

$$\overset{\mathcal{T}}{\mathcal{T}}\ A(0)$$
$$\dfrac{\overline{\mathcal{T}}\ A(t)}{\mathcal{T}'\ A(a)}$$
$$\overline{\overline{\mathcal{T}}}\ A(a')$$

is a derivable rule in Ψ.

Proof. For $\mathcal{T}$ = T, we have

$$T\ A(0)$$

$$F\ A(t)$$

$$\text{CUT}$$

$$T\ A(0) \supset A(t) \qquad\qquad F\ A(0) \supset A(t)$$
$$\beta \qquad\qquad\qquad\qquad\qquad\qquad CI$$
$$F\ A(0) \qquad T\ A(t) \qquad F\ A(a) \supset A(a')$$
$$\qquad\qquad\qquad\qquad\qquad\alpha$$
$$\$_b \qquad\qquad \$_b \qquad\quad T\ A(a)$$
$$\qquad\qquad\qquad\qquad\qquad\alpha$$
$$\qquad\qquad\qquad\qquad\quad F\ A(a')$$

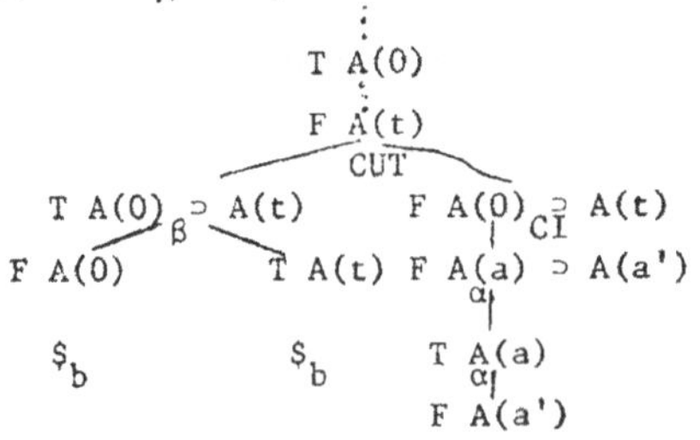

The two leftmost branches can be extended to closed subtrees.

For $\mathcal{Y}$ = F, we have

$$F\ A(0)$$
$$T\ A(t)$$
$$CUT$$

$T\ \neg A(0) \supset \neg A(t)$	$F\ \neg A(0) \supset \neg A(t)$
β	CI
	$F\ \neg A(a) \supset \neg A(a')$

$F\ \neg A(0)$ $T\ \neg A(t)$

α α α

$T\ A(0)$ $F\ A(t)$ $T\ \neg A(a)$

$\$_b$ $\$_b$

α

$F\ \neg A(a')$

α

$F\ A(a)$

α

$T\ A(a')$

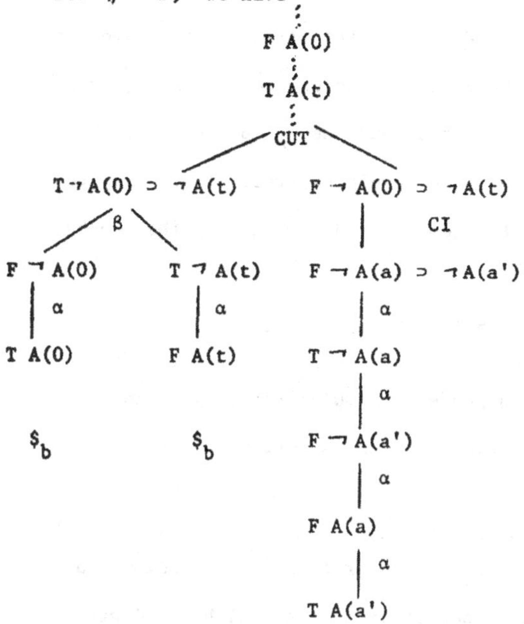

Again the two leftmost branches can be extended to closed subtrees.

<u>Proposition 10</u>. Let S be a set of formulas, X(b) be a formula containing a parameter b such that b does not occur in any formula of S, t be an arbitrary term. Then if S ∪ {X(b)} has a closed tableau $\mathcal{T}$ in Ψ, $\mathcal{T}$ can be transformed into a closed tableau for S ∪ {X(t)} in which the same rules are used as in $\mathcal{T}$.

Proof. If we merely substitute t for b throughout $\mathcal{T}$ we may not have all rules still in force, because the restriction on the δ or complete induction rules may not always be met. We thus consider the finite set of parameters $\{a_i\}$, $i = 1, \ldots, n$, that 1) occur in t, 2) are different from b, and 3) introduced in $\mathcal{T}$ by a δ or complete induction rule. We take a set of distinct parameters $\{b_i\}$, $i = 1, \ldots, n$, where each b_i occurs neither in $\mathcal{T}$ nor t, and we replace each a_i throughout the $\mathcal{T}$ by b_i. We then replace b by t throughout the resulting tree. It is easy to see that the final tree satisfies the requirements of the proposition.

Similarly to Proposition 10, we can prove:

Proposition 11. If you can prove a formula $P(b)$, you can prove the formula $P(t)$ where b is a parameter and t is an arbitrary term.

Thus whenever we prove a particular formula, we know we can prove all substitution instances of it. This will be used without mention in the application of the axiom and theorem rules and the derived closure conditions a and d, i.e., when a formula TP is introduced into a proof by an axiom or theorem rule or a branch is closed by a formula FP, and reference is made to a given arithmetically false formula or a formula already known provable, the formula referred to may be one from which P results by the substitution of terms for free variables.

We say two tableaux are structurally equal when there exists a 1-1 correspondence between the formulas such that corresponding formulas are either both end formulas of a branch or follow by the same rules from parents that correspond and have predecessors that correspond. (We note that structural equality will always imply equal ranks under the ordinal assignments we will use.)

We call a tableau compatible with a given string of symbols when no free variable occurring in the string is introduced by a δ or complete induction rule in the tableau. Similarly to proposition 10 we have:

Proposition 12. Any subtree of a tableau for a set S can be transformed into and replaced by a structurally equal subtree comaptible with any given string, the result being again a tableau for S.

This completes our description and discussion of the fundamental properties of Ψ.

§2. A Constructive Consistency Proof for Ψ

We now present a version of a well-known consistency proof due to Gentzen [2]. We will show that not every formula is provable in Ψ, that in particular no false numerical formula is provable in Ψ. We will do this by showing that, given a proof of a numerical formula A, one can transform it in finitely many steps to a proof of A using only α, β and cut rules. It is then easy to see that a formula with such a proof can only be true.

Before beginning the proof, however, we must make a number of definitions and observations.

Let $\mathcal{T}$ be a closed tableau for a formula X that may contain applications of the derived replacement rule, in addition to the rules of Ψ. We say such a $\mathcal{T}$ is a normal derivation of X if it contains no applications of the repetition rule. Clearly a formula X is provable in Ψ iff it has a normal derivation, and in this paragraph we will also sometimes call a normal derivation for X a proof of X. The formula being proved will always be the origin of such a tableau.

In order to prove the consistency of the system Ψ inductively we associate with every normal derivation a specific ordinal number. For this we need several supplementary concepts.

We define the <u>degree</u> of an induction inference $\frac{F\ A(0) \supset A(t)}{F\ A(a) \supset A(a')}$ to be the degree of A(a) - recall that the degree of a cut inference is the degree of the cut formula.

The <u>depth</u> of a formula X in a normal derivation is the maximal degree of the cut inferences and complete induction inferences with a conclusion not below X in the normal derivation. If there is no such cut or induction inference, X has depth 0. By the definition, the first formula of a normal derivation always has depth 0. As one proceeds down a derivation tree the depth of a successor can change only if that successor is the result of a cut or induction inference, in which case the depth may or may not increase but will certainly not decrease. Thus the depth of a formula X is actually a certain measure of the complexity of the part of the proof ending with X (and its cut formula, in the case of a cut). This measure will be used in an essential way in the induction. It will be the thing that "makes the proof work."

We now associate with every formula of a normal derivation in Ψ a specific ordinal, to be called the <u>Ψ-rank</u> of the formula:

1) Every end formula of a branch has Ψ-rank 1.

2) The predecessor of a replacement rule conclusion has the same Ψ-rank as the conclusion.

3) If the conclusion of an α, γ or δ rule has the Ψ-rank γ_1, the rank of its predecessor is $\gamma_1 \oplus 1$.

4) If the conclusions of a β rule have Ψ-ranks, γ_1 and γ_2 respectively, their predecessor is $\gamma_1 \# \gamma_2$.

5) If the conclusions of a cut rule have Ψ-ranks γ_1 and γ_2 and n is the difference between the depth of the conclusions and the depth of their predecessor, the predecessor has Ψ-rank $\omega_n(\gamma_1 \# \gamma_2)$.

6) If the conclusion of a complete induction rule has Ψ-rank γ and n is the difference between the depth of the conclusion and the depth of its predecessor, the predecessor has Ψ-rank $\omega_n(\gamma \odot \omega)$.

Throughout our discussion of the consistency of Ψ in this paragraph we will use "rank" to mean Ψ-rank. We will drop the circles around the ordinal functions throughout the rest of this work, however, for we will not often have occasion to refer to the corresponding arithmetic functions.

The rank of a normal derivation is defined to be the rank of its first formula. This rank is always smaller than the first ϵ-number ϵ_0. For by property 15 of the ordinal functions discussed in the introduction, one sees that in every case the predecessor of a formula has a rank $< \epsilon_0$ when the ranks of the successors are $< \epsilon_0$. Since the end formulas have rank $1 < \epsilon_0$, the first formula of any derivation in Ψ also has a rank $< \epsilon$

Proposition 13. Let $\mathcal{T}$ be a finite tree with no complete induction conclusion that is a subtree of two other trees $\mathcal{T}'$ and $\mathcal{T}''$ in such a way that the only points of $\mathcal{T}'$ and $\mathcal{T}''$ not in $\mathcal{T}$ occur below end points of $\mathcal{T}$. If one or more end points of $\mathcal{T}$ have lower ranks in $\mathcal{T}''$ than in $\mathcal{T}'$, and no end points have a higher rank in $\mathcal{T}''$ than in $\mathcal{T}'$, then all points above or equal to end points of different ranks (in particular, the origin) have lower ranks in $\mathcal{T}''$ than in $\mathcal{T}'$ (and other points have the same ranks).

Proof. We will prove the proposition inductively for all the points of $\mathcal{T}$. Let X be a point of $\mathcal{T}$ and assume the proposition true for all points below X in $\mathcal{T}$ if there are any. If X is an end point of $\mathcal{T}$ the proposition is true by assumption.

If X is not an end point, we consider individual cases, noting that X is above an end point that has a different rank in $\mathcal{T}'$ and $\mathcal{T}''$ iff at least one of its successors is either such an end point or lies above one. If X is not above such an end point (it cannot be equal to one), neither are any of its successors. Since by the induction hypothesis all the successors of X then have the same ranks in $\mathcal{T}'$ and $\mathcal{T}''$, so will X in the two trees, since the rank of X is computed by the same rule from the ranks of its successors in the two trees. Thus let us assume now that X lies above an end point of differing ranks, so that at least one of its successors is equal to or lies above such an end point.

Case 1. Let X have a successor Y that is the result of an α,γ, or δ rule. Let the rank of Y in $\mathcal{T}''$ be γ'', the rank of Y in $\mathcal{T}'$ be γ'. By assumption $\gamma'' < \gamma'$. Then the rank of X in $\mathcal{T}''$ is $\gamma'' + 1$, while the rank of X in $\mathcal{T}'$ is $\gamma' + 1$. But $\gamma'' + 1 < \gamma' + 1$ since $\gamma'' < \gamma'$ (this is not a case of one of the monotonicity properties mentioned in the introduction but is easily shown to follow from them), which is the desired result.

Case 2. Let X have successors Y_1, Y_2 that are the results of a β rule or a cut rule. Let the ranks of Y_1, Y_2 in $\mathcal{T}''$ be respectively γ_1'', γ_2'', and in $\mathcal{T}'$ be γ_1', γ_2'. Let Y_1 be a successor of X that is equal to or above an end point of differing rank (Y_2 may also be such a successor, but it doesn't matter). By assumption

$$\gamma_1'' < \gamma_1'$$

$$\gamma_2'' \leq \gamma_2'.$$

The rank of X in $\mathcal{T}''$ is

$$\gamma_1''\#\gamma_2'' \quad \text{or} \quad \omega_n(\gamma_1''\#\gamma_2'')$$

(for the β and cut rule respectively) while the rank of X in $\mathcal{T}'$ is

$$\gamma_1'\#\gamma_2' \quad \text{or} \quad \omega_n(\gamma_1'\#\gamma_2')$$

respectively. But then

$$\gamma_1''\#\gamma_2'' < \gamma_1'\#\gamma_2' \quad \text{and}$$

$$\omega_n(\gamma_1''\#\gamma_2'') < \omega_n(\gamma_1'\#\gamma_2')$$

by the commutativity and strict monotonicity of $\#$ (properties 13c and 13e) and the strict exponent monotonicity of $\alpha_n(\beta)$ (property 7d)

Thus the rank of X in $\mathcal{T}''$ is less than the rank of X in $\mathcal{T}'$ in both cases.

<u>Proposition 14</u>. Let $\mathcal{T}$ be a finite tableau in Ψ in which only $\alpha, \beta, \gamma, \delta$ and replacement rules are used. The rank of $\mathcal{T}$ is then a finite ordinal n (here $n = \underbrace{1 \# \ldots \# 1}_{n \text{ summands}} = \underbrace{1 + \ldots + 1}_{n \text{ summands}}$).

This proposition can be proved in a manner similar to that used in the proof of proposition 13.

We now turn to the consistency proof. Using in addition to strictly finite means only transfinite induction up to the first ϵ-number, we will show:

<u>Theorem 1</u>. No false numerical formula is provable in Ψ.

This will imply:

<u>Theorem 2</u>. The system Ψ is syntactically consistent.

For assume you could prove both A and $\neg A$ in Ψ with tableaux $\mathcal{T}_1$ and $\mathcal{T}_2$. Then you could prove $0 = 0'$ with the tableau.

$$\dfrac{F\ 0 = 0'}{\begin{array}{cc} T \neg A & F \neg A \\ FA & \mathcal{T}_2 \\ \mathcal{T}_1. & \end{array}} \ \text{CUT}$$

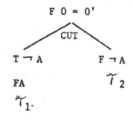

But $0 = 0'$ is a false numerical formula, so such a proof is impossible.

Now let us turn to the proof of Theorem 1. Let P be an arbitrary provable numerical formula and let us be given a particular normal derivation for it. (As noted before, this derivation must have rank $< \epsilon_0$.

We will show how we can obtain from the normal derivation of P a normal derivation of P such that:

(1) only α, β, replacement, and cut rules are used;

(2) all the formulas in the tableau are numerical;

(3) every branch contains a numerically false atomic formula.

The following lemma will then give us the theorem:

Lemma 1. If there exists a normal derivation for a formula P satisfying conditions (1), (2) and (3) immediately above, P is true.

Proof. Let $\mathcal{T}^P$ be a (finite) tableau for a formula P satisfying (1) and (2). We show that if P is false (so that F P is true) one can find a branch of $\mathcal{T}^P$ in which every formula is true. So let us assume P is false and prove this assertion by complete induction on the level of a formula in the tableau, i.e., let us prove that for any level for which there exist formulas in the tableau at that level, there is a path θ containing precisely the formulas on some branch up to that level and such that all the formulas on it are true. Since there are formulas of only finitely many levels in our proof, this will imply that there is a complete branch containing only true formulas.

F P is the only formula of level 1 and it is true by assumption. Now assume the assertion true for a level ℓ. Let T_ℓ be the subtree of $\mathcal{T}^P$ containing all the formulas of $\mathcal{T}^P$ of level $\leq \ell$. We wish to show that $T_{\ell+1}$ contains a branch with only true formulas if T_ℓ does. We differentiate the following cases:

1) There are no points of level $\ell + 1$ as successors of end points of branches in T_ℓ that contained only true formulas in T_ℓ.

In this case the subtree T_ℓ, and hence $T_{\ell+1}$, contains a complete branch with only true formulas and we are done.

2) There is a path θ of T_ℓ that contains only true formulas in T_ℓ and whose end formula in T_ℓ has a successor or successors in $T_{\ell+1}$. We must distinguish between the rules used to obtain the successors.

2a) The rule used is an α rule. By the truth definition an α is true iff both α_1 and α_2 are true. So the extended branch must contain only true formulas.

2b) The rule used is a β rule, so that θ is extended to two branches θ_1 and θ_2 by the adjunction of β_1 and β_2. By the truth definition, β is true iff at least one of β_1 and β_2 is true. So at least one of θ_1, θ_2 contains only true formulas.

2c) The rule used is the cut rule, so again θ is extended to two branches θ_1 and θ_2 by the adjunction of X and $\overline{X}$ for some (numerical) formula X. It can be shown inductively that X and $\overline{X}$ always have opposite truth values, so at least one of them is true. Again at least one of θ_1, θ_2 contains only true formulas.

2d) The rule used in extending θ is a replacement rule. It can be shown by induction that if X results from Y by a replacement rule application, X and Y have the same truth value. So the new formula is also true and we are done.

So we wish to transform our arbitrary normal derivation $\mathcal{T}$ of P to one satisfying conditions (1), (2), and (3). To show how we can do this, we use the concept of a *quantifier set*:

If a normal derivation contains a cut whose cut formula contains a universal or existential subformula Q that is not within another such subformula - a so-called maximal quantified subformula - we define the *quantifier set* S_Q associated with Q by the following two conditions:

1. The occurrences of the formula Q in both the formulas of the cut belong to the quantifier set S_Q.

2. If a formula G has direct descendents by an α or β rule and has a component Q^* belonging to the quantifier set S_Q, then the parts of the direct descendents that correspond to Q^* also belong to S_Q.

Note that quantifier set members are unsigned formulas that are occurrences of (sub)formulas in the bodies of certain signed formulas in a proof. We do not wish to write out a formal definition of the part corresponding to Q* after the application of an α or β rule to a formula containing Q*. That part is, briefly, the complete image of Q*, if there is one, in the direct descendents. Thus, if Q* is a subformula of an α or β (it cannot be all of an α or β) it will have an image in one of the two successors and the image will be the same formula. In the case of a replacement rule, the image of Q* may differ from Q by having different terms. We see:

Q1) A quantifier set contains either only existential formulas or only universal formulas.

Q2) Only formulas that occur below the cut associated with the quantifier set contain elements of the quantifier set, and no formula can contain more than one such element.

Q3) If in a tableau using only α, β, replacement and cut rules one replaces all the elements of a quantifier set by the same unsigned formula (note that this replacement takes place inside formulas) all the rule applications retain their characters. (This is easy to see since all the members of the quantifier set are occurrences of existential or universal formulas.)

We will show that we can obtain from our arbitrary normal derivation $\mathcal{T}$ of P a tableau $\mathcal{T}'$ for P satisfying the following conditions.

(1') only α, β, replacement and cut rules are used in $\mathcal{T}'$;

(2') no parameters or predicate variables occur in any formula of $\mathcal{T}'$;

(3') there are no formula occurrences πQ_1 and $\bar{\pi} Q_2$ in $\mathcal{T}'$ such that Q_1 and Q_2 are in the same quantifier set.

(4') every branch is either closed or contains a γ or δ formula.

Then the following lemma combined with lemma 1 will give us the theorem:

Lemma 2. If there exists a tableau $\mathcal{T}'$ for a numerical formula P satisfying conditions (1'), (2'), (3'), and (4') immediately above, one can find a tableau $\mathcal{T}''$ for P satisfying conditions (1), (2), and (3) of lemma 1.

Proof. We first note that since P is numerical, each maximal quantified subformula of the form $(\forall x)A$ or $(\exists x)A$ occurring in formulas of $\mathcal{T}'$ belong to a (single) quantifier set. We replace all such subformulas by numerical formulas as follows: let S_Q be a quantifier set; if F Q' appears in $\mathcal{T}'$ for a $Q' \in S_Q$, we replace every member of S_Q by $0 = 0$, and otherwise we replace every member of S_Q by $0 = 0'$. Let the resulting tree be called $\mathcal{T}''$. By Q3, all the rules of $\mathcal{T}'$ retain their characters, so $\mathcal{T}'$ is a tableau satisfying condition (1). Since no parameters or predicate variables occurred in $\mathcal{T}'$ (so that none occur in $\mathcal{T}''$ either) and all the formulas in $\mathcal{T}''$ are also quantifier-free, they must be numerical. Thus $\mathcal{T}''$ satisfies condition (2). Now we wish to see that $\mathcal{T}''$ satisfies condition (3), i.e., that every branch contains a numerically false atomic formula. Let θ be a branch of $\mathcal{T}'$. By condition (4') θ is either closed or contain a γ or δ formula.

First assume the branch θ is closed due to a closure set S. Since the formulas of S are atomic and may contain no parameters or predicate variables, they must be numerical. Clearly, no matter which closure condition was appropriate, one of the formulas in the closure set must be false. And under the replacement rule these formulas appear unchanged in $\mathcal{T}''$, so that the branch in $\mathcal{T}''$ corresponding to θ also contains a false atomic formula. Now consider a branch θ that contains a γ or δ formula $\pi(qx)A$, where q is $\forall$ or $\exists$. Let us also denote this formula by X. As noted at the beginning of the proof, $(qx)A$ must belong to some quantifier set S_Q. Now if there is a $Q' \in S_Q$ such that $F\,Q'$ apears in $\mathcal{T}'$, then π must also be F by condition (3') on $\mathcal{T}'$ and $(qx)A$ is replaced by $0 = 0$ in $\mathcal{T}''$, so that the branch corresponding to θ in $\mathcal{T}''$ contains $F\ 0 = 0$, a false atomic formula. If there is no $Q' \in S_Q$ such that $F\,Q'$ appears in $\mathcal{T}'$, in particular π is not F, i.e., is T, and $(qx)A$ is replaced by $0 = 0'$, so that in this case the branch corresponding to θ in $\mathcal{T}''$ contains $T\ 0 = 0'$, again a false atomic formula. This completes the proof of lemma 2.

Thus we now need only show how to obtain from an arbitrary normal derivation $\mathcal{T}$ of a numerical formula P a normal derivation of P satisfying conditions (1'), (2') (3') and (4') of lemma 2. We first introduce some further concepts.

The <u>beginning part</u> of a normal derivation consists of the part of each branch down to and not including the conclusion of the first application of a γ,δ, or complete induction rule on the branch. We call the partial branches of the beginning part of a proof the <u>beginning branches</u> of the proof. The <u>final formula</u> of a beginning branch is defined to be: 1) the last formula on the branch, if the original branch had no γ,δ, or complete induction rule applied; 2) the γ,δ, or complete induction rule predecessor (not premise) which occurs at the end of the beginning branch. Thus for any normal derivation:

a) the formula being proved is in the beginning part.

b) only α,β, replacement and cut rules appear in the beginning part.

c) the final formulas of the beginning part are either end formulas of the original proof or predecessors of γ, δ, or complete induction rule conclusions.

We now describe a series of operations that may be performed on a normal derivation of a numerical formula. One operation will be such that it cannot be performed more than once without some other operation intervening, and performance of the other operations will result in a tableau of lower rank; thus the repeated application of these operations must terminate. Each operation assumes that the previously described operations are not applicable.

Let $\mathcal{T}$ be a normal derivation for a numerical formula P.

<u>Operation 1</u>. If some parameter is introduced <u>in the beginning part of</u> $\mathcal{T}$, we replace every such parameter by the numeral

0 throughout the proof. When this is done the first formula remains unchanged because, as a numerical formula, it contains no parameters. It is clear that all rules and closure conditions still apply.

Operation 2 - eliminating induction inferences. Let us assume that some final formula of the beginning part is the predecessor of an induction inference

$$\frac{F\ A(0) \supset A(t)}{F\ A(a) \supset A(a')}$$

so that the part of the derivation we consider has the form

	ranks
θ	
$F\ A(0) \supset A(t)$	
θ'	
Y	$\omega_n(\sigma \cdot \omega)$
$F\ A(a) \supset A(a')$	σ
$\mathcal{T}$	

Here θ and θ' are parts of the path P_Y, $\mathcal{T}$ is the part of the proof below $F\ A(A) \supset A(a')$, and Y is the predecessor of the conclusion of the rule, being possibly the same as $F\ A(0) \supset A(t)$. We let σ be the rank of $F\ A(a) \supset A(a')$ in the given derivation, so that Y has the rank $\omega_n(\sigma \cdot \omega)$, where n is the difference between the depths of Y and $F\ A(a) \supset A(a')$.

The term t that occurs in the premise in the beginning part contains no parameter (by Operation 1); it is thus numerical and has a specific numerical value $0^{(k)}$. (The integer k denotes the number of apostrophes attached to 0.)

We will eliminate the induction inference in question, replacing it with other inferences.

I. If the term t has the numerical value 0, $A(0)$ and $A(t)$ are equivalent formulas, and by two applications of the α rule to $F\ A(0) \supset A(t)$, we have

$$\begin{array}{c} \theta \\ F\ A(0) \supset A(t) \\ \theta' \\ Y \\ T\ A(0) \\ F\ A(t) \end{array}$$

By proposition 2, we can extend this branch to a closed subtree in which only $\alpha, \beta, \gamma, \delta$, and replacement rules are used. By proposition 14, the rank of $F\ A(t)$ will be a finite ordinal m. Then the rank of Y will be $m + 2$. Since $1 \leq \sigma$,

$$m + 2 = 1 \cdot (m + 2) \leq \sigma \cdot (m+2) < \sigma \cdot \omega \leq \omega_n(\sigma \cdot \omega).$$

The predecessor of the induction inference conclusion thus has a smaller rank than it did in the given normal derivation.

II. If t has a numerical value $0^{(k)}$ with $k \neq 0$, for every numeral $z < 0^{(k)}$,[4] we modify $\mathcal{T}$ to a $\mathcal{T}_z$ so that

$$\begin{array}{c} \theta \\ F\ A(0) \supset A(t) \\ \theta' \\ Y \\ F\ A(z) \supset A(z') \\ \mathcal{T}'_z \end{array}$$

[4] This is an arithmetical $<$.

is a closed subtree. This is possible by proposition 10. Note that this may not be a derivation since $F\ A(z) \supset A(z')$ is not necessarily the conclusion of any rule. In every such subtree $F\ A(z) \supset A(z')$ has the rank σ.

a. When $k = 1$ one obtains from $F\ A(0) \supset A(t)$ the formula $F\ A(0) \supset A(0')$ by a replacement inference. So the tree we have <u>does</u> form part of a derivation and the rank of Y there is σ, which is smaller than $\omega_n(\sigma \cdot \omega)$, the rank of the same formula in the original derivation.

b. When $k > 1$ we obtain a normal derivation by changing the part of the derivation we are working with as follows:

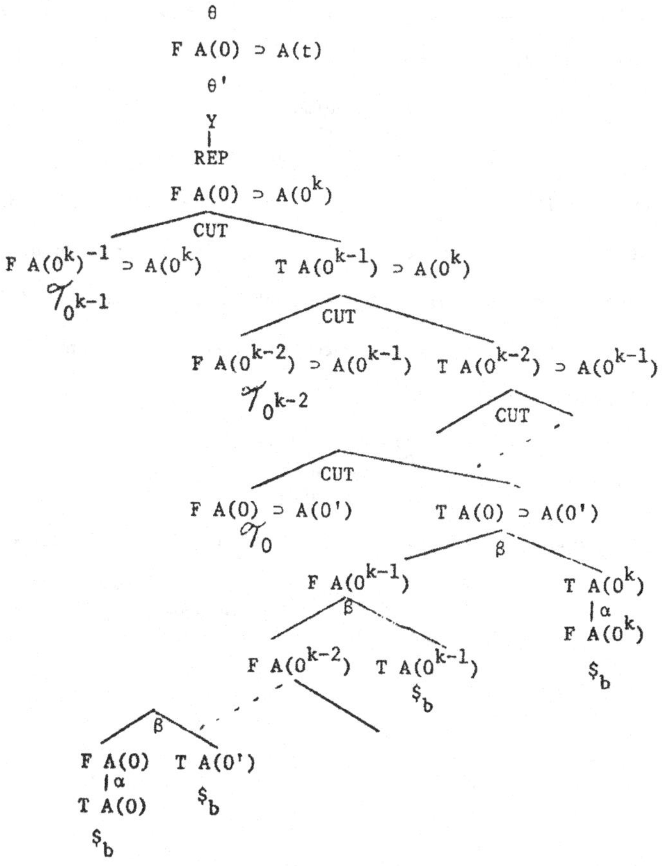

By proposition 2 (and the obvious fact that one can eliminate repetitions from proofs) we can again extend every branch with a $\$_b$ beneath it to a closed subtree using only $\alpha, \beta, \gamma, \delta$, and replacement rules.

Let us now calculate the rank of Y in the transformed proof. By proposition 14, $T\ A(0) \supset A(0')$ has a finite rank $m < \omega$. Except for the first cut denoted, each cut has above it a cut of the same degree, so that no depth change can occur between a cut formula and its predecessor except at the first cut. In particular the rank of the predecessor of the lowest cut formulas will be $\omega_0(\sigma \# m) = \sigma \# m$. Working up similarly, we find that the rank of the predecessor $T\ A(0^{k-1}) \supset A(0^k)$ of the next to highest cut denoted is

$$\underbrace{\sigma \# \ldots \# \sigma}_{k-1 \text{ summands}} \# m.$$

Then the difference in depth at the highest cut formula denoted must be the same difference n as in the original induction inference. We thus calculate the rank of the formula $F\ A(0) \supset A(0^k)$, and thus of Y, since the former formula is the conclusion of a replacement rule, to be

$$\omega_n(\underbrace{\sigma \# \ldots \# \sigma}_{k \text{ summands}} \# m).$$

By properties 13e and 13f of the ordinals (see the introduction)

$$\underbrace{\sigma \# \ldots \# \sigma}_{k \text{ summands}} \# m \leq \underbrace{\sigma \# \ldots \# \sigma}_{k+m \text{ summands}} < \sigma \cdot \omega$$

so $\omega_n(\sigma\#\sigma\# \ldots \#\sigma\#m) < \omega_n(\sigma \cdot \omega)$ and the order of $F\ A(0) \supset A(0^{(k)})$ in the resulting derivation is less than the order of $F\ A(a) \supset A(a')$ in the original derivation.

Operation 2 consists in the removing of an induction inference and replacing the part of the derivation containing it with new parts as specified in I or II. As we have seen, in doing this we lower the order of the predecessor of the inference's conclusion (Y above) in every case. From proposition 13 (the beginning part has no CI inferences) it follows that the first formula FP of the normal derivation also has a smaller rank.

Operation 3 - eliminating conjugate quantificational rules. Let us assume that the beginning part has among its final formulas the predecessors of γ and δ inferences whose bodies belong to the same quantifier set.

Thus the derivation has a section

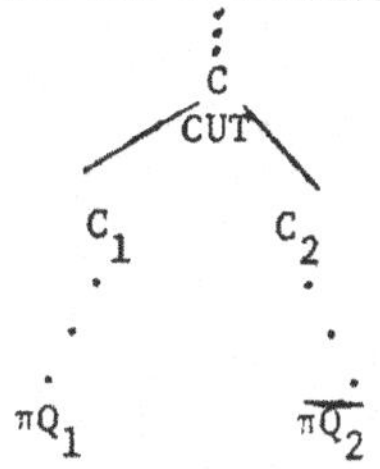

where C_1 and C_2 are cut formulas of the cut the quantifier set is associated with, πQ_1 is a γ, $\overline{\pi Q_2}$ a δ, and Q_1 can be transformed to Q_2 with the use of replacements. The degree of the cut in question is greater than 0, so C_1 and C_2 both have a depth greater than 0. Then there is in the proof a lowest formula D above C_1 and C_2 that has a smaller depth than C_1. D may be equal to C or may be above C. In any case D is the predecessor of a cut (since we are in the beginning part of a derivation). Thus D_1 and D_2, the cut formulas succeeding D have a degree d that is greater than or equal to the degree c of C_1 and C_2, while all cuts above D have a degree lower than d.

So the normal derivation has a section of the form

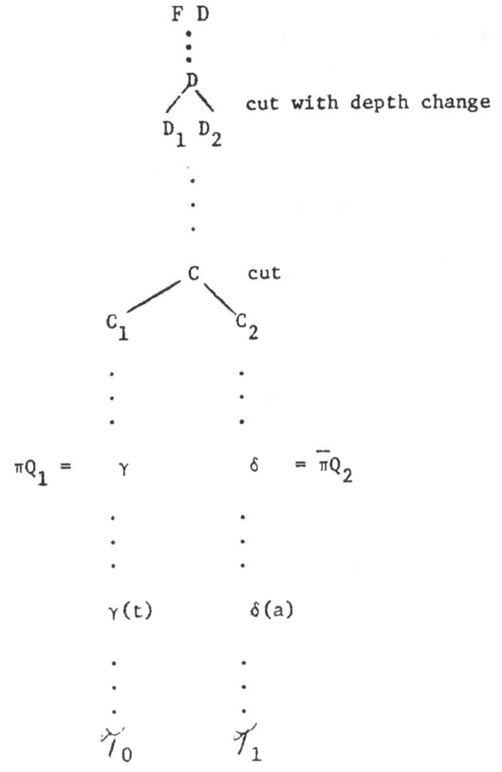

We transform this derivation as follows:

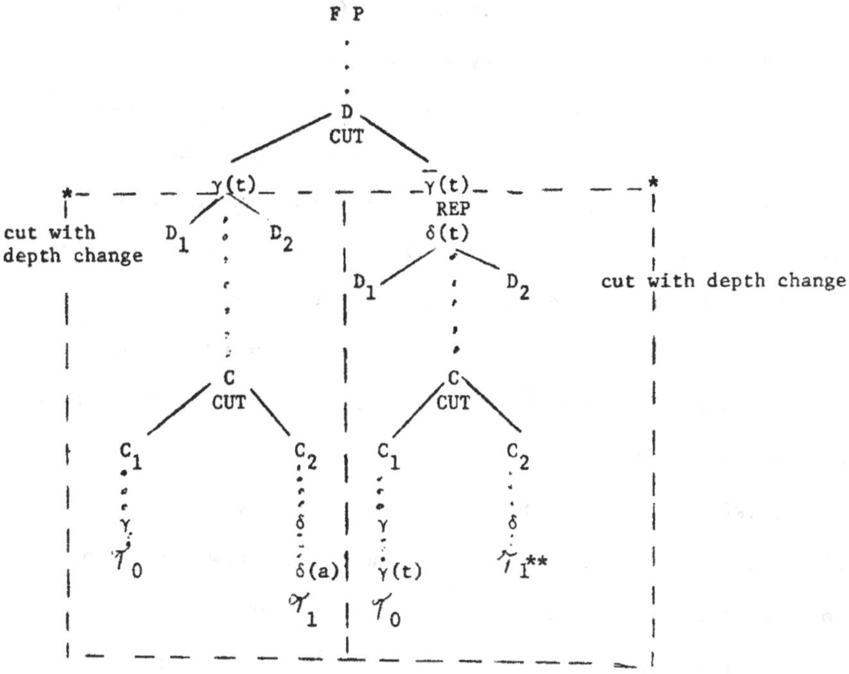

Here we have introduced a new cut after D with cut formula Q_1. We have followed $\overline{\gamma}(t)$ by a replacement inference to obtain $\delta(t)$. The rest of the proof tree is structurally the same except outside the beginning part except for changes of parameters. The starred left section below $\gamma(t)$ is the result of modifying the part corresponding to it in the original proof so that it is compatible with $\gamma(t)$. This can be done by proposition 12. The same is true for the significance of the starring of the right section. The subtree $\mathcal{T}_1^{**}$ of the right section is the result of replacing a by t in $\mathcal{T}_1$. The new subtree consisting of $\mathcal{T}_1^{**}$ and all the formulas above it still closes by proposition 10. The sections we have changed are also a valid part of a proof, since $\gamma(t)$ and $\overline{\gamma}(t)$, the only questionable formulas are obtained by legal rules.

Looking forward to the rank calculation, we note that the depths of D, D_1, D_2, C_1, C_2 are all the same in the new derivation as in the old one. Thus if we let n be the difference in depths that occurred between D and D_1 (or D_2 or C_1 or C_2):

$$n = \text{depth}\ (D_1) - \text{depth}\ (D)\ ,$$

and let m be the difference in depths between D and $\gamma(t)$ (or $\overline{\gamma}(t)$) in the new derivation

$$m = \text{depth}\ (D_1) - \text{depth}(\gamma(t))\ ,$$

then also

$$n - m = \text{depth}\ (\gamma(t)) - \text{depth}\ (D)\ .$$

Here m must be greater than zero since the depth of D_1 is d, the degree of D_1, and the depth of $\gamma(t)$ is less than d (for the degree of $\gamma(t)$ is less than c, the degree of $C_1 = \gamma$, where $c \leq d$, and all the other cuts above $\gamma(t)$ also have degrees $< d$ by the assumption of the occurrence of a depth change at D in the original derivation).

We now calculate the rank of this new normal derivation. The parts of the derivation under the $\gamma(t)$ and $\delta(t)$ introduced by the new cut are like parts of the original derivation except that in both cases the conclusion of a γ or δ rule is missing on one branch. By proposition 13, the formulas $\gamma(t)$ and $\delta(t)$ have lower ranks than D originally did. Let σ be the natural sum of the ranks of the successors of D in the original normal derivation. Then in the new derivation the natural sums of the ranks of the successors of $\gamma(t)$ and $\delta(t)$ are ordinals σ_1 and σ_2, both of which are smaller than σ. The ranks of $\gamma(t)$ and $\delta(t)$ [and thus of $\overline{\gamma}(t)$] are $\omega_m(\sigma_1)$ and $\omega_m(\sigma_2)$ respectively.

In the new derivation the formula D has the order

$$\omega_{n-m}(\omega_m(\sigma_1)\#\omega_m(\sigma_2)) < \omega_{n-m}(\omega_m(\sigma))$$

$$= \omega_n(\sigma)$$

(by 13g).

Now consider our given tableau $\mathcal{T}$ for P. Let us also call it $\mathcal{T}_1$ and denote its rank by σ_1. If any of the operations 1-3 are applicable to $\mathcal{T}$ we transform $\mathcal{T}$ according to the following procedure:

Assume for some $i \geq 1$ we have a tableau $\mathcal{T}_i$ for P of rank σ_i to which some operation is applicable. Apply the first of the operations that is applicable and call the result $\mathcal{T}_{i+1}$, its rank σ_{i+1}.

We can see that such a procedure must terminate because any infinite sequence $\sigma_1, \sigma_2, \ldots$ we could generate in this manner would contain a decreasing subsequence (since operation 1 cannot be applicable to two successive $\mathcal{T}_i$ and the two other operations yield a tableau of smaller rank), but no decreasing sequence of ordinals can exist. Thus we must finally obtain a tableau $\mathcal{T}_i$ for P to which none of the operations is applicable. Such a tableau satisifes the following properties:

(a) the beginning part of $\mathcal{T}_i$ contains no parameters;

(b) there are no formula occurrences πQ_1 and $\bar{\pi} Q_2$ in the beginning part of $\mathcal{T}_i$ such that Q_1 and Q_2 are in the same quantifier set;

(c) every final formula of the beginning part is either the end formula of a closed branch or is the predecessor of a γ or δ rule.

Now let $\mathcal{P}'$ be the beginning part of $\mathcal{P}_i$. We can now easily verify that $\mathcal{P}'$ satisfies conditions (1'), (2'), (3') and (4') of lemma 2. For only α, β, replacement and cut rules are used in beginning parts, so condition (1') is satisfied and the properties (a), (b), and (c) yield respectively conditions (2'), (3') and (4').

This completes the proof of Theorem 1.

§3. The Infinitary System Ω; Ψ as a subsystem of Ω

In §5 we will see that the system Ψ contains a formula $A(a)$ with the properties:

a) For every numeral z, $A(z)$ is a provable formula in Ψ.

b) The universal formula $(\forall x)A(x)$ is not provable in Ψ.

This shows a certain incompleteness of the system Ψ that is called ω-incompleteness. In this part of the paper an extension Ω of the system Ψ will be studied. The extended system Ω is ω-complete in a certain constructive sense. A classical equivalent of Ω will be seen to be complete in a stronger sense.

Let us now describe Ω.

The primitive symbols of the system Ω are the same as those in Ψ except that individual parameters are not used.

The <u>terms</u> and <u>formulas</u> of Ω are defined in the same way as in Ψ except that parameters are not used.

Tableaux in Ω are constructed like tableaux in Ψ, i.e., they are trees, but in Ω we will see that the trees may contain denumerably many points. Any closed branch, however, can be shortened to a finite closed branch, so that we may assume that in any closed tableau all the branches are of finite length.

The <u>rules</u> of Ω are the same as those of Ψ except that we replace (1) the δ rule of Ψ by the δ_ω rule: if δ is an element of P_Y, we may adjoin $\delta(0)$, $\delta(0')$, ...$\delta(z)$,..., as the first second,..., $z+1^{st}$,..., successors of Y and (2) the γ rule by the γ' rule: if γ is on P_Y, then we may adjoin as the sole successor of Y the formula $\gamma(z)$, where z is any numeral. The closure conditions of Ω are the same as those of Ψ, but may be stated in a simpler form because no parameters occur.

Thus we can visualize the proof-theoretic structure of Ω as follows:

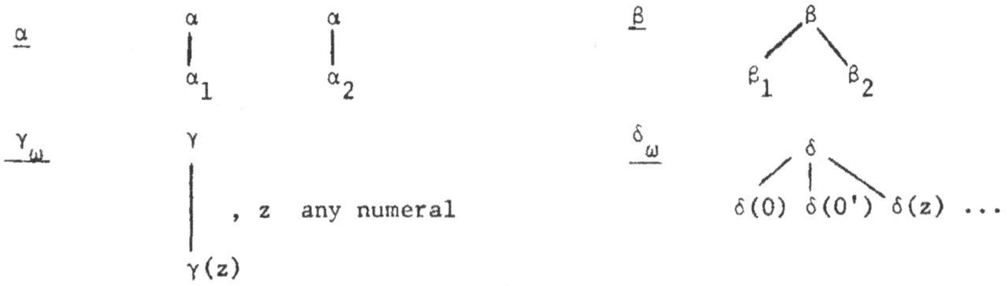

<u>Cut</u>

$$\begin{array}{ccc} & X & \\ \swarrow & & \searrow \\ Y & & \overline{Y} \end{array}$$

<u>Assumption</u>

$$\begin{array}{l} s_1 \\ \cdot \\ \cdot \quad s_i \epsilon S,\ i = 1,\ldots,n \\ \cdot \\ s_n \end{array}$$

<u>closure</u>: A branch is <u>closed</u> if a) it contains a false atomic numerical formula; b) it contains atomic formulas $A, \overline{A'}$, where A is equivalent to A'.[5]

Of course, if, in the role of constructive mathematicians, we wish to present an infinite proof of Ω, we must give a method of <u>constructing</u> such a proof. The following is an example of a proof in the system Ω[6]:

We note that

$$F(\forall y)(\exists x)(y\cdot y' = x + x)$$

$F(\exists x)(0\cdot 0' = x + x)$	$F(\exists x)(0'\cdot 0'' = x + x)$	...	$F(\exists x)(z\cdot z' = x + x)$ ...
$F\ 0\cdot 0' = 0 + 0$	$F\ 0'\cdot 0'' = 0' + 0'$		$F\ z\cdot z' = t_z + t_z$

where t_z is defined as follows:

---

[5] Here as in Ψ we need closure condition b) only because we include predicate variables in Ω. Closure condition a) is more simply stated here because any arithmetically unsatisfiable set of atomic formulas without parameters consists only of numerical formulas and thus must contain at least one false one, which itself alone forms an arithmetically unsatisfiable set. We no longer need closure condition c) - that a branch is closed if it contains formulas $T\ s = t, A, \overline{A'}$, where A and A' are atomic and equivalent relative to $s = t$ - because in such a case the branch would be closed according to condition a) or b). For all terms are numerical in Ω, so that if $s = t$ were false, the first closure condition would hold, while if $s = t$ were true, A and A' would be equivalent and the second closure condition would hold.

[6] I have used the same examples as Schütte for similar systems so that one can also compare proofs in his systems to proofs in our systems.

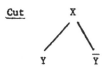

$t_0 = 0,$

$t_{z'} = t_z + z'.$

The essential difference about derivations in Ψ and Ω is illustrated by the examples given, which are both proofs of the same formula. The derivation in Ω appears simpler than the derivation in Ψ. It requires, however, that a suitable term t_z be given for every numeral z. The two formulas on lines 3, 14 and 15 of the derivation in Ψ show us how to find the t_z. Thus, a primitive derivation in Ω generally requires the calculation of certain terms, which is not, as in the system Ψ, expressed explicitly in the derivation. In Ω, rather, metalogical considerations are put forth. In any case, the calculations of terms must be defined constructively. If a derivation in Ψ is given for the formula, one can deduce from the derivation a constructive procedure for the term calculation.

Definitions of proofs, derived rules, derived closure conditions, etc. are made for Ω as they were for Ψ. We can easily prove propositions 1-6 for Ω also.

We note that the system Ω is ω-complete in a constructive sense, i.e., when it can be shown constructively that for all z, $P(z)$ has a proof, then the formula $(\forall x)P(x)$ has a proof. For assume that we know by a constructive proof that $P(z)$ is provable in Ω for every numeral z. This means that we know there are closed tableaux T_z for each of the signed formulas $F\ P(z)$. So we can form the tableau

$$
\begin{array}{c}
F\ (\forall x)P(x) \\
T_0 \quad T_{0'} \quad \ldots \quad T_z \quad \ldots
\end{array}
$$

which is a proof of $(\forall x)P(x)$.

Now let us have a diversion into classical mathematics to see that this system is <u>complete</u> in the usual classical sense. The proof we give is almost identical to that of the Gödel completeness theorem for first order logic.

By a <u>formation tree</u> is meant a tree in which each end point is atomic, and such that for every other point one of the following conditions holds:

1) The point is of the form $(A \wedge B)$, $(A \vee B)$, or $(A \supset B)$, and has A and B for its first and second successors, and it has no other successors, or the point is of the form $\neg A$ and it has A for its sole successor.

2) The point is of the form $(\forall x)A$ or $(\exists x)A$ and it has $A^x_0, A^x_{0'}, \ldots, A^x_z, \ldots$ as its successors.

We call $\mathcal{T}$ a formation tree <u>for A</u> if $\mathcal{T}$ is a formation tree with A at the origin. We define the subformulas of A to be the formulas which occur in the formation tree for A. (It can be shown that any formula has only one formation tree.)

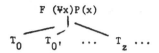

Let E be all the well-formed unsigned formulas of Ω. By an <u>interpretation</u> I of E over the domain of natural numbers is meant a function which assigns to each n-place predicate variable P an n-place number-theoretic relation P^*, and to each n-place predicate constant Q and function constant f the number theoretic predicate Q^* or function f^* to which they are supposed to correspond.

All terms in Ω are numerical and have values as described in §1 of part I. If the terms $t_1, \ldots, t_n$ have the numerical values $z_1, \ldots, z_n$ and P is an n-place predicate variable or constant, the atomic formula $Pt_1,\ldots,t_n$ is called <u>true (false) under I</u> iff the n-tuple $\langle z_1,\ldots,z_n\rangle$ stands (does not stand) in the relation P^*. By the valuation tree $\mathcal{T}^*$ for A with respect to the interpretation I, we mean the formation tree $\mathcal{T}$ for A together with an assignment of truth values to all the points of $\mathcal{T}$ such that the truth values of the end points (which are atomic formulas) are determined by I as above and the truth values of all the other points are determined from the truth values of their immediate successors in the tree by the conditions:

B_1. The formula $\neg X$ receives the value t if X receives the value f and f if X receives the value t.

B_2. The formula $(X\wedge Y)$ receives the value t if X and Y both receive the value t, and otherwise $(X\wedge Y)$ receives the value f.

B_3. The formula $(X\vee Y)$ receives the value t if at least one of X and Y receives the value t, and otherwise $(X\vee Y)$ receives the value f.

B_4. The formula $(X\supset Y)$ receives the value f if X and Y receive the respective values t and f, and otherwise $(X\supset Y)$ receives the value t.

F_1. The formula $(\forall x)A$ receives the value t if for every numeral z, A_z^x receives the value t, and otherwise $(\forall x)A$ receives the value f.

F_2. The formula $(\exists x)A$ receives the value t if for at least one numeral z, A_z^x receives the value t, and otherwise $(\exists x)A$ receives the value f.

We note that although every formula can be shown classically to be either true or false under an interpretation, there may be some formulas that cannot be shown constructively to have either truth value. For example, consider a formula of the form $(\forall x)A$. If we assume that every A_z^x has a truth value, we may not conclude that it is constructively true that either all these truth values are t or that not all these truth values are t. We may not be able to prove either statement.

Trivial Propositions. 1) Equivalent formulas have the same value with respect to any given proposition. 2) A numerical formula has the same value with respect to any interpretation over the domain of natural numbers as its value by the definition in §1.

We can now define the truth value of a signed formula in terms of the truth value of its unsigned body as follows: the value of a signed formula with prefix T is the same as the truth value of its body, while the truth value of a signed formula with prefix F is the opposite of that of its body.

A formula A (signed or unsigned) is called satisfiable in the domain of the natural numbers if it is true in at least one interpretation over the domain of the natural numbers. More generally, a set S of formulas (signed or unsigned) is called <u>satisfiable</u> (or <u>simultaneously satisfiable</u>) if there is at least one interpretation under which all the elements of S are true. A formula A is called <u>valid</u> in the domain of the natural numbers if it is true under every interpretation over the domain of natural numbers. For brevity, we will speak simply of satisfiability and validity.

It is a trivial exercise to verify that under any interpretation over the natural numbers,

N_1. α is true iff α_1 and α_2 are both true.

N_2. β is true iff at least one of β_1 and β_2 is true.

N_3. γ is true iff $\gamma(z)$ is true for every numeral z.

N_4. δ is true iff $\delta(z)$ is true for at least one numeral z.

As a consequence of the above facts, we have the following laws concerning satisfiability. Here S is any set of (signed) formulas.

S_1. If S is satisfiable and $\alpha \epsilon S$, then $\{S, \alpha_1, \alpha_2\}$ is satisfiable.

S_2. If S is satisfiable and $\beta \epsilon S$, then at least one of the two sets $\{S, \beta_1\}$, $\{S, \beta_2\}$ is satisfiable.

S_3. If S is satisfiable and $\gamma \epsilon S$, then for every numeral z the set $\{S, \gamma(z)\}$ is satisfiable.

S_4. If S is satisfiable and $\delta \epsilon S$, then there exists a numeral z such that the set $\{S, \delta(z)\}$ is satisfiable.

Since we are using tableaux with signed formulas, it will be convenient to use this characterization of the requirements on the truth values and satisfiability.

One can show just as in first order logic (cf. Smullyan [1], p. 55):

Theorem 3. Every formula provable in Ω is valid (semantic consistency).

(There are two non-constructive points in the classical proof by contradiction of this proposition that is usually given. The first involves assuming that any formula is either true or false, but could be remedied. The crucial one involves assuming for a certain statement $P(x)$ that either $(\forall x)P(x)$ or $(\exists x)\neg P(x)$, is true, which is not constructively valid.)

Definition. By a Hintikka set over the natural numbers we mean a set S of signed formulas of Ω such that the following conditions hold:

H_0a. S does not contain any atomic numerically false formula.

H_0b. S does not contain two formulas A and $\overline{A'}$ where A and A' are atomic and equivalent.

H_1. If $\alpha \epsilon S$, then α_1 and α_2 are both in S.

H_2. If $\beta \epsilon S$, then $\beta_1 \epsilon S$ or $\beta_2 \epsilon S$.

H_3. If $\gamma \epsilon S$, then for every numeral z, $\gamma(z) \epsilon S$.

H_4. If $\delta \epsilon S$, then for at least one numeral z, $\delta(z) \epsilon S$.

Hintikka's lemma for first order number theory. Every Hintikka set over the natural numbers is satisfiable over the natural numbers.

Proof. We must find an interpretation over the natural numbers in which all the elements of S are true. It suffices to say what predicates are assigned to the predicate variables occurring in S. We assign to an n-place predicate variable P the relation P that contains precisely those n-tuples $\langle z_1,\ldots,z_n\rangle$ such that some $T\ Pt_1,\ldots,t_n$ is in S where t_i is equal in value to z_i for each i. We must show that each element X of S is true under this interpretation. We do this by induction on the degree of X. If X is of degree 0 and its predicate is a predicate constant, X is true by H_0a. If X is of degree 0 and its predicate is a predicate variable, X is true by the assignment just defined and the fact that H_0b holds. Now suppose that X is of positive degree and that every element of S of lower degree is true. Since X is not of degree 0, it is either some α, β, γ or δ formula. If it is an α formula, by H_1 both α_1 and α_2 are in S, by the induction hypothesis both of these are true, and so by N_1, S is true. The remaining cases are equally trivial.

Systematic tableaux in Ω. To prove completeness we must show that there is a proof, i.e., a closed tableau in Ω, for every valid formula of Ω. To do this we provide a procedure (one out of many possible ones) that will lead to a proof whenever one exists.

The systematic tableau for a signed formula X is the tableau "constructed" as follows: At the first stage X is placed at the origin. Now consider the $n+1$-st stage. If the tableau is closed, we stop. Otherwise for each open branch θ we consider the set T_{n+1}^{θ} (which may be empty) defined as follows: for $1 \leq i \leq n+1$ we define Y_i^{θ} to be the formula of level i that is on the branch θ if there is one on the branch that is non-atomic -- if there is no such formula there is no formula Y_i. T_{n+1}^{θ} is the set of all the Y_i^{θ}. (At any stage there can only be finitely many members in each Y_i^{θ} set.) Now we successively take each member Y_i^{θ} of T_{n+1}^{θ} (in the order of their levels, say) and do the following for it: we take every open branch θ^* containing θ and

1) If Y is an α, we extend θ^* to the branch $(\theta^*, \alpha_1, \alpha_2)$;

2) If Y is a β, we simultaneously extend θ^* to the two branches (θ^*, β_1) and (θ^*, β_2);

3) If Y is a δ, we simultaneously extend θ^* to the denumerable set of branches $(\theta^*, \delta(0)), (\theta^*, \delta(0')) \ldots$.

4) If Y is a γ, we take the first numeral z such that $\gamma(z)$ does not occur on θ^* (if any) and we extend θ^* to $(\theta^*, \gamma(z))$.

The resulting tableau may contain only branches that are finite in length, or may have open branches that are infinite in length.

It is easy to see that:

Proposition 15. In the systematic tableau for any formula, every open branch is a Hintikka set over the natural numbers.

From Proposition 15 and Hintikka's lemma for first order number theory, we have:

Proposition 16. In the systematic tableau $\mathcal{T}$ for any formula, every open branch is simultaneously satisfiable over the natural numbers.

This yields:

Theorem 4 (Completeness of Ω). If X is valid over the natural numbers, X is provable in Ω.

Proof. Suppose X is valid. Let $\mathcal{T}$ be the finished systematic tableau starting with FX. If $\mathcal{T}$ contained an open branch θ, then by Proposition 16, θ would be simultaneously satisfiable. Hence FX, being an element of θ, would be satisfiable, contrary to the hypothesis that X is valid. Thus X is provable and its systematic tableau provides a proof. (This proof is also not constructive, even if X is known to be true constructively, for we cannot exhibit the false formula on each branch of the systematic tableau for X.)

We will now see in what sense Ψ is a subsystem of Ω.

We first introduce a system Ψ' equivalent with Ψ that will be easier to compare with Ω. The system Ψ' is defined just like Ψ with the following exceptions:

1) the rule of complete induction $\frac{F\ A(0) \supset A(t)}{F\ A(a) \supset A(a')}$ for an a new to the branch

is not allowed in Ψ'.

2) the following closure condition is added: I) a branch is closed if it contains a formula $F(A(0) \wedge (\forall x)(A(x) \supset A(x'))) \supset A(t)$ for any formula $A(a)$ and any term t.

Let us now show the equivalence of the systems Ψ and Ψ'.

Proposition 17. Every formula derivable in Ψ' is also derivable in Ψ.

Proof. Let P be a formula provable in Ψ'. By definition we then have a closed tableau for it constructed and closing according to the rules of Ψ'. But all the rules of Ψ' are also rules of Ψ, so the construction of the tableau is allowed in Ψ. As is, it may not close in Ψ, however, for some branches may close according to condition I, which is not a closure condition in Ψ. We extend each such branch to a closed branch as follows. Let such a branch contain the formula

$$F(A(0) \wedge (\forall x)(A(x) \supset A(x'))) \supset A(t).$$

We add on to the end of the branch

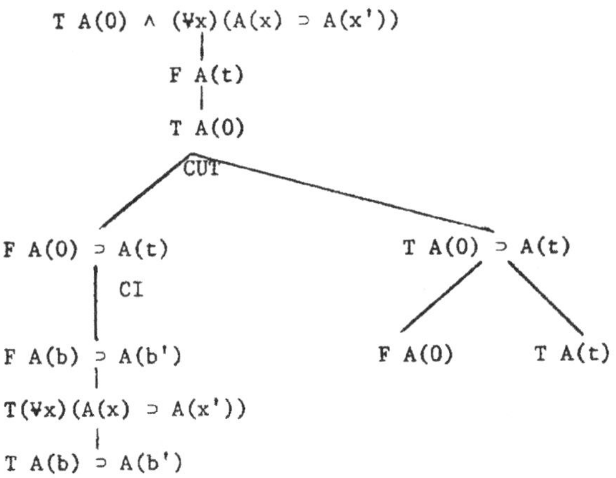

where b is a parameter not occurring in the branch. The tableau resulting is still a tableau in Ψ and each of its branches is either closed or contains some formula and its conjugate. By Proposition 2 of §1 each branch is either closed or can be extended to a closed branch.

Proposition 18. Every formula derivable in Ψ is derivable in Ψ'.

Proof. Let P be a formula provable in Ψ. We can thus assume we have a closed tableau for P constructed and closing according to the rules of Ψ. The closure rules of Ψ' include the closure rules of Ψ so all branches are closed in terms of Ψ'. If the complete induction rule is applied in the derivation, the tree is not yet a tableau in Ψ'. We replace each use of the complete induction rule by the use of rules of Ψ' as follows:

Let us assume that $F\ A(b) \supset A(b')$ was concluded from $F\ A(0) \supset A(t)$. At the place where $F\ A(b) \supset A(b')$ occurs we insert the tree

$$T\ A(0)$$

$$F\ A(t)$$

CUT

$T\ (A(0) \wedge (\forall x)(A(x) \supset A(x'))) \supset A(t)$ $\qquad$ $F(A(0) \wedge (\forall x)(A(x) \supset A(x'))) \supset A(t)$

$F\ A(0) \wedge (\forall x)(A(x) \supset A(x'))$ $\qquad$ $T\ A(t)$ $\qquad$ #

$F\ A(0)$ $\qquad$ $F(\forall x)(A(x) \supset A(x'))$ $\qquad$ \$

\$ $\qquad$ $F\ A(b) \supset A(b')$

Here the branch marked with a # sign is closed by closure condition I in Ψ', while the branches marked with a \$ sign can both be extended to closed branches, again by Proposition 2 of §1 which also holds for Ψ', since Ψ' has the necessary α, β, γ, δ, and repetition rules and its closure conditions include those of Ψ.

Thus we have:

Proposition 19. A formula is provable in Ψ iff it is provable in Ψ'.

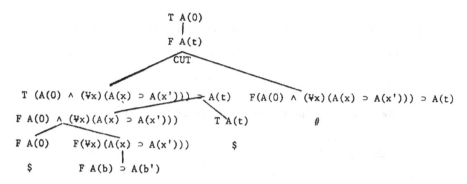

Now we will prove that Ψ is a subsystem of Ω in the sense that all universal closures of any formula provable in Ψ are provable in Ω (a universal closure of a formula with parameters is the result of replacing the different parameters with distinct variables and then adding universal quantifiers for the variables at the front of the formula.) By the last proposition of the previous section, it is sufficient to prove that the system Ψ' is a subsystem of Ω in the same sense.

Theorem 5. Any closed formula derivable in Ψ' with a proof of degree n is provable in Ω with a tableau of degree n and of rank $\leq \omega + i < \omega + \omega = \omega \cdot 2$ for some integer i.[7]

We prove this theorem with the use of two lemmas.

Lemma 1. A tableau $\mathcal{T}$ for a closed formula X constructed and closed according to the rules of Ψ' can be transformed into a tableau $\mathcal{T}'$ for X such that 1) $\mathcal{T}$ and $\mathcal{T}'$ have the same rank, 2) only the rules of Ω are used in $\mathcal{T}'$, 3) every branch in $\mathcal{T}'$ can be obtained from a branch of $\mathcal{T}$ by making an appropriate substitution of numerals for terms throughout the branch, 4) all branches in $\mathcal{T}'$ are closed according to the closure conditions of Ψ'.

Proof. Let a tableau $\mathcal{T}$ be given in Ψ' for a closed formula X. We first substitute a numeral, say 0, throughout $\mathcal{T}$ for all the parameters first introduced by a cut or γ rule. This clearly preserves rank and rule applications and closures. The only parameters that remain in the tree were introduced into the tree by δ rule applications. Let us consider a branch

---

[7] From now on we will be speaking of the rank as defined in the introduction. The Ψ-rank will not be used again.

of the form

$$\begin{array}{c} \vdots \\ \delta_1(a_1) \\ \vdots \\ \delta_2(a_2) \\ \vdots \\ \delta_{n-1}(a_{n-1}) \\ \vdots \\ Y \\ \delta_n(a_n) \\ A \end{array}$$

where A is the ordered set of trees under $\delta_n(a_n)$ and no δ rule applications are made in A. We will successively eliminate the parameters as follows. We first form the tree

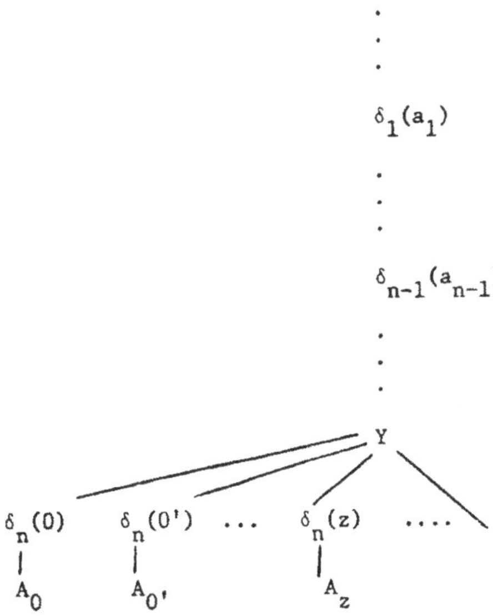

where A_z is the result of substituting z for a_n throughout A. We note this second transformation does not change the rank of the tableau, for the superformula of the δ rule conclusion we are changing does not change its rank. Also, in doing this all the rules and closures of Ψ' are preserved except that we have replaced δ rule applications by δ_ω rule applications. We repeat this procedure for each parameter on the branch. We are working at all times with a tree in which there are only finitely many levels, so this is a finite process and when we are done, clearly no individual parameters exist on the branch. Now we consider any γ rule application on the branch where $\gamma(t)$ is concluded. Here t must be a numerical term, so let z be the numeral with the same value. We then replace all occurrences of t resulting from the t's in $\gamma(t)$ by z. All closures are retained and *the rank of the tableau is not changed*. The rules used in the tableau are now the α, β, γ', δ_ω, and cut rules. It is clear that the tableau we have formed satisfies the requirements for $\mathcal{T}'$ in the statement of the lemma.

Lemma 2. A tableau $\mathcal{T}'$ of degree n possessing the properties 2) and 4) of Lemma 1 can be extended to a tableau $\mathcal{T}''$ in Ω of degree n that is closed according to the closure conditions of Ω, and this extension can be made by adding only subtrees of certain finite ranks $m_1,\ldots,m_i,\ldots$ below end points of $\mathcal{T}'$.

Proof. Let $\mathcal{T}'$ be a tableau possessing the properties 2) and 4) of Lemma 1. If we can show that all the branches of $\mathcal{T}'$ are either closed in Ω or can be extended to closed branches in Ω by using rules of Ω we will be done. We consider the following cases for a branch:

a) The branch closes according to a closure condition of Ψ' that is also a closure condition of Ψ. We have already noted that the closure conditions of Ψ for Ω tableaux are included in the Ω closure conditions, so that the branch is also closed in Ω.

b) The branch closes according to condition I of Ψ'. Let $F(A(0) \wedge (\forall x)(A(x) \supset A(x'))) \supset A(t)$ be the formula causing the branch to close. Here t must have a numerical value $0^{(k)}$, by which we mean 0 followed by k apostrophes. We extend this branch as follows:

$T\ A(0) \wedge (\forall x)(A(x) \supset A(x'))$

$F\ A(t)$

$T\ A(0)$

$T(\forall x)(A(x) \supset A(x'))$

$T\ A(0) \supset A(0')$

$T\ A(0') \supset A(0'')$

.
.
.

$T\ A(0^{(k-1)}) \supset A(0^{(k)})$

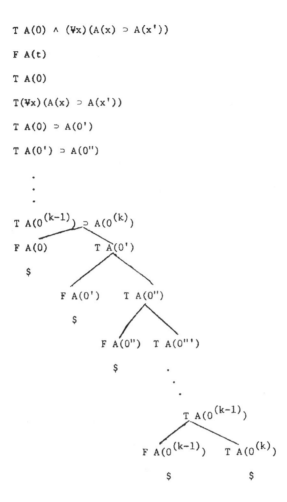

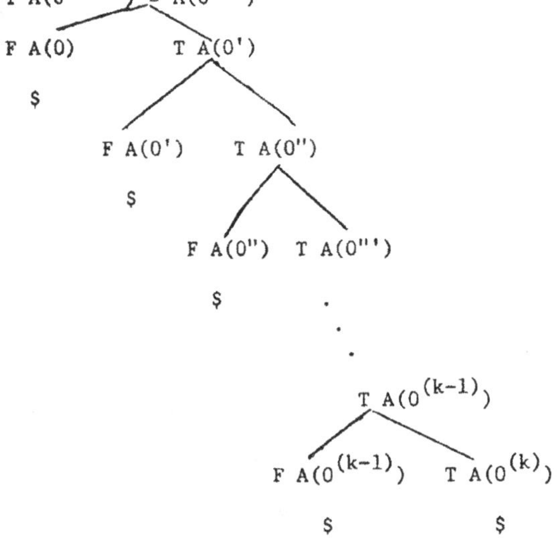

Here of course we don't go beyond the third formula in the tree if $k = 0$. By proposition 2 (which we mentioned could also be proven for Ω) all the branches resulting (the end points of which are marked with a \$ sign) can be extended to closed branches in Ω using trees of rank $< \omega$ (without use of the cut rule).[8] This means that the rank of the closed tree that we have added to the end of the branch that closed by rule I is finite.

Proof of Theorem 5. Let a closed formula X be provable in Ψ' with a tableau $\mathcal{T}$. We transform this tableau to a proof of X (possibly with cuts) in Ω with the use of Lemma 1 and Lemma 2. The intermediate tree has the same order as $\mathcal{T}$, which is an integer i. By adding on the subtrees to get the proof in Ω we raised to ranks of the end points to ranks $\leq m_j < \omega$. Thus by the rank Lemma 2, the rank of the resulting tree is $\leq \omega + i$. This number is less than $\omega + \omega$.

<u>Proposition 20</u>. Any universal closure of a formula provable in Ψ' is provable in Ω by a cut-bounded tableau of rank $\leq \omega + \omega$.

Proof. Let R be a formula provable in Ψ' and let R' be a universal closure of it. We can start out a derivation of R' in Ω by successive applications of the δ_ω rule to get rid of the quantifiers added to form the universal closure. The result of this is a tree with end points all the possible combinations of substitutions of numerals for parameters in F R. (Here the formula R is always preceded by the prefix F.) It thus suffices to find tableaux in Ω for these substitution

---

8 If m_i is the degree of A_i, and A_i and $\overline{A_i}$ are equivalent, $\{A_i, \overline{A_i}\}$ has a closed tableau in Ω of rank $\leq 2m_i$.

instances of R. Let $\mathcal{T}$ be the tableau of Ψ' that proves R. If we make a particular substitution for the parameters in R throughout $\mathcal{T}$, we get a proof of the substitution instance in Ψ' (cf. proposition 11). Let the substitution instances be enumerated in some order $R_1, R_2, \ldots,$ and let the Ψ' proofs of them be $\mathcal{T}_1, \mathcal{T}_2, \ldots$. Then we apply Proposition 19 to the $\mathcal{T}_i$ to obtain proofs $\mathcal{T}_i''$ in Ω for the substitution instance R_i all of which are of a rank $\leq \omega + j$ for some integer j and of the same finite degree. We now form a proof for R' as follows:

$$F\ (\forall x_1)(\forall x_2)\ \ldots\ (\forall x_k)R(x_1, x_2, \ldots, x_k)$$

$$\delta_\omega$$

$$F\ (\forall x_2)\ \ldots\ (\forall x_k)R(z_1, x_2, \ldots, x_k)$$

$$\vdots$$

$$\mathcal{T}_i''$$

(Here $F\ R_i$ is the first formula in $\mathcal{T}_i''$.)
We can see that the rank of this cut-bounded proof is $\leq \omega + j + k < \omega + \omega$.

This completes the proof of the proposition.

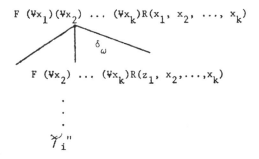

We note that proposition 20 implies that if Ψ is syntactically inconsistent, so is Ω. For assume there were a formula $A(a_1,\ldots,a_n)$ in Ψ such that both it and its negation were provable in Ψ (and thus in Ψ'). By Proposition 11, one could also prove in Ψ both $A(0,\ldots 0)$ and its negation. But these are closed formulas. Thus by Proposition 20, one could prove both $A(0,\ldots 0)$ and its negation in Ω (with cut-bounded tableaux).

§4. A Constructive Consistency Proof for Ω

Proposition 21. Every closed tableau in Ω of degree 0 and rank $\leq \mu$ can be transformed into a cut-free closed tableau with the same origin branch and with a rank $\leq 2^\mu$.

Proof. Let there be given a closed tableau T of degree 0 and rank $\leq \mu$, and let the origin branch contain successively the formulas $Y_1, \ldots, Y_n$. We take for an induction hypothesis that the proposition holds for all tableaux of degree 0 and rank $< \mu$. If Y_n has no successors, the proposition is trivally true. Otherwise let formulas X_i be the immediate successors of Y_n in T and let the formula X_i have the ranks $\leq \mu_i$. The formulas X_i are themselves at the tops of subtrees S_i. Let T_i be the tree

$$\begin{array}{c} Y_1 \\ \cdot \\ \cdot \\ \cdot \\ Y_n \\ S_i \end{array}$$

It is easy to see that this can be viewed as a closed tableau with origin branch $Y_1, \ldots, Y_n, X_i$, degree 0 and rank $\leq \mu_i < \mu$.

By the induction hypothesis these tableaux can be transformed into cut-free closed tableaux

$$\begin{array}{c} Y_1 \\ \cdot \\ \cdot \\ \cdot \\ Y_n \\ S_i{*} \end{array}$$

with the same origin branches and with ranks $\leq 2^\mu$. We have two cases to consider.

1) Let the rule used to obtain the X_i not be a cut. Then we form a new tree by putting the $S_i{*}$ beneath the initial formulas $Y_1, \ldots, Y_n$, using the same rule as in the original tree to obtain the formulas X_i (which are at the top of the $S_i{*}$). By the rank Lemma 1, the rank of the resulting cut-free tree is $\leq 2^\mu$.

2) Let the rule used to obtain the X_i be a cut (so $i = 2$). Then we form a new tree as follows. We take $S_1{*}$, delete X_1 from the top of it, and place the resulting subtree(s) below $Y_1, \ldots, Y_n$. We then take $S_2{*}$, delete X_2 from the top of it, and place the resulting subtree(s) below the end point of every branch of the tree we had just formed using $S_1{*}$. We claim that the resulting tree is a closed cut-free tableau with origin branch $Y_1, \ldots, Y_n$ and the required rank. It is clearly cut-free.

Now since X_1 and X_2 were of degree 0, no rules could be applied to them, so that all the other formulas of S_1* and S_2* had to be obtained from $Y_1, \ldots, Y_n$ or their descendents. Thus all old rule application still retain their characters in the new tree. We next wish to show that all the branches are closed. Let us be given such a branch. It must consist of the string $Y_1, \ldots, Y_n$ followed by a branch B_1 of S_1* (minus X_1) and then by a branch B_2 of S_2* (minus X_2) below it. And the branches of S_1* and S_2* were all closed when preceded by $Y_1, \ldots, Y_n$. Thus if the branch $Y_1, \ldots, Y_n, X_1, B_1$ closed without need of X_1, the branch in question in the new tree is also closed. It is also closed if $Y_1, \ldots, Y_n, X_2, B_2$ closed without need of X_2. We thus need to consider only the case when both branches needed the cut formulas to close. We consider the possible cases:

i) Let X_1 close its branch by closure condition a, so that X_1 is a false atomic (signed) formula. Then X_2, being a <u>true</u> signed formula could only close its branch by closure condition b. This means that $Y_1, \ldots, Y_n, B_2$ contains a formula that has a body equivalent to that of X_2 and a prefix opposite to that of X_2. Such a formula, having a body equivalent to that of X_1, and having the <u>same</u> prefix as X_1, will be a false signed atomic formula, as X_1 was. Thus this formula could be used to close the branch $Y_1, \ldots, Y_n, X_2, B_2$, contrary to the assumption that X_2 was needed.

ii) Let X_1 close its branch by closure condition b, so that

$Y_1, \ldots, Y_n, B_1$ contains a formula U that has a body equivalent to that of X_1 and a prefix opposite to that of X_1. The case when X_2 closes its branch by closure condition a is like case 1) above. So it remains to consider the case when X_2 also closes its branch by closure condition b. In this case, $Y_1, \ldots, Y_n, B_2$ contains a formula W with a body equivalent to that of X_2 and a prefix opposite to that of X_2. The prefix of W is thus opposite to that of U, and since equivalence is a transitive relation, the body of U is equivalent to that of W. Hence since U and W are both contained in the branch $Y_1, \ldots, Y_n, B_1, B_2$, this branch must close by closure condition b.

We still have to show that the new tree has the proper rank. The tree

$$\begin{array}{c} Y_1 \\ \cdot \\ \cdot \\ \cdot \\ Y_n \\ S_1{*} - X_2 \end{array}$$

clearly has rank $\leq 2^{\mu_1}$. When we add on the X_2-deleted subtrees of $S_2{*}$ to its end points, we effectively raise the ranks of its end points to at most 2^{μ_2}. By the rank Lemma 2, this implies that the rank of the resulting tree is

$$\leq 2^{\mu_2} + 2^{\mu_1} \leq 2^{\max(\mu_1,\mu_2)} + 2^{\max(\mu_1,\mu_2)} = 2^{\max(\mu_1,\mu_2)} \cdot 2 =$$

$$2^{\max(\mu_1,\mu_2) + 1} = 2^{\mu}.$$

This completes the proof of proposition 21.

Proposition 22. Every closed cut-bounded tableau in Ω of degree $n > 0$ and rank $\leq \mu$ can be transformed into a closed tableau with the same origin branch that is of degree $< n$ and rank $\leq 2^{\mu}$.

Proof. Let there be given a closed tableau of degree n and rank $\leq \mu$, and let the origin branch in the tree contain $Y_1, \ldots, Y_n$. We take as an induction hypothesis that the proposition holds for all tableaux of degree $\leq n$ and of rank $< \mu$. Let formulas X_i be the immediate successors of Y_n in the tree and let the formula X_i have rank $\leq \mu_i$. The formulas X_i are at the tops of trees S_i themselves. Let T_i be the tree

$$\begin{array}{c} Y_1 \\ \cdot \\ \cdot \\ \cdot \\ Y_n \\ S_i. \end{array}$$

It is easy to see that these can be viewed as closed tableaux with origin branches $Y_1, \ldots, Y_n, X_i$ that are of degree $\leq n$ and of rank $\leq \mu_i < \mu$. By the induction hypothesis these trees can be transformed into closed tableaux of degree $< n$.

$$Y_1$$

$$\vdots$$

$$Y_n$$

$$S_i^*$$

with the same origin branches and with ranks $\leq 2^{\mu_i}$. We have two cases to consider.

1) Let the rule used to obtain the X_i not be a cut of degree n. Then we form a new tree by putting the S_i^* beneath the formulas $Y_1, \ldots, Y_n$, using the same rule as in the original tree to obtain the formulas X_i (which are at the tops of the S_i^*). By the rank Lemma 1, the rank of the new closed cut-free normal tree is $\leq 2^{\mu}$ and the tableau is of degree $< n$.

2) Let the rule used to obtain the X_i be a cut of degree $n > 0$. If $\mu_1 = \mu_2 = 0$, i.e., neither of the cut formulas have successors, we can just remove the cut since, not being atomic, neither of the cut formulas can be used to close a branch. We consider the following two subcases for $\mu_1 > 0$ or $\mu_2 > 0$.

a. One of the cut formulas is an α formula and the other, $\bar{\alpha}$, is a β formula.

We thus have a tree of the following form:

$$Y_1$$

$$\vdots$$

$$Y_n$$

$$\alpha \qquad \bar{\alpha}$$

$$A \qquad\quad B$$

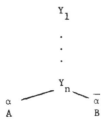

where α over A is, say, $S_1{*}$ and $\overline{\alpha}$ over B is $S_2{*}$. (Thus A and B are both ordered sets of trees.)

We transform this tree into the following tree:

(*)

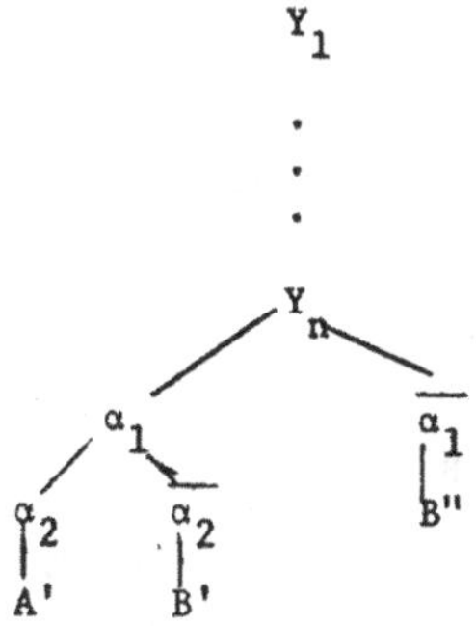

where here A' is the result of deleting occurrences of α_1 and α_2 in A that were obtained from α; B' (B'') is the result of eliminating any formula in B that occurs below an $\overline{\alpha_1}$ ($\overline{\alpha_2}$), and then eliminating all the remaining $\overline{\alpha_1}$'s and $\overline{\alpha_2}$'s.

We claim that the tree (*) is a closed tableau with the same origin branch as the original tableau and with cuts only of degree $< n$ and with a rank $\leq 2^{\mu}$.

We leave it to the reader to verify that (*) is a closed tableau with the same origin branch as the original tree and with a degree $< n$. We now verify that its rank is correct. It is clear that the ranks of α_2, $\overline{\alpha_2}$, and $\overline{\alpha_1}$ are respectively $\leq 2^{\mu_1}$, $\leq 2^{\mu_2}$, and $\leq 2^{\mu_2}$. Let η be the maximum of μ_1 and μ_2. Then the rank of α_1 is $\leq 2^{\eta} + 1$ and the rank of Y_n, which is the rank of the tableau is $\leq 2^n + 2^n \leq 2^n + 2^n = 2^{n+1} = 2^\mu$.

b. One of the cut formulas is a γ and the other, $\overline{\gamma}$, is a δ formula.

Thus we have a tree of the following form

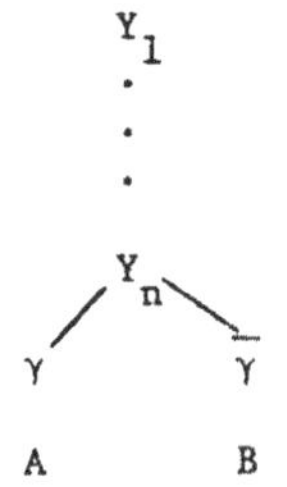

where $\begin{matrix}\gamma \\ A\end{matrix}$ is, say, $S_1{}^*$ and $\begin{matrix}\overline{\gamma} \\ B\end{matrix}$ is $S_2{}^*$, A and B being ordered sets of trees.

We replace this tree by a tree of the form

$$(**)\qquad \begin{matrix} Y_1 \\ \cdot \\ \cdot \\ \cdot \\ Y_n \\ C \end{matrix}$$

where C is formed as follows. Each of the points $\gamma(z)$ in A that is derived from γ are replaced by a cut of the form

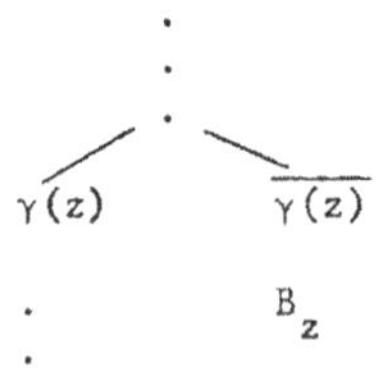

where B_z is the tree that results from B when, for each $z' \neq z$, each subtree starting with a $\overline{\gamma(z')}$ (coming from $\overline{\gamma}$ by a δ^{ω} rule application) is eliminated and also all occurrences of $\overline{\gamma(z)}$ are eliminated.

We leave it to the reader to verify that (**) is a closed tableau with the same origin branch as the original tableau and with a degree $< n$. We notice that if, in the tableau

$$\begin{array}{c} Y_1 \\ \vdots \\ Y_n \\ A \end{array}$$

(whose rank is $\leq 2^{\mu_1}$) we replace all the occurrences of $\gamma(z)$ that we changed to get C simply by cuts of the form

$$\begin{array}{ccc} & \vdots & \\ \gamma(z) & & \overline{\gamma(z)} \\ \vdots & & \end{array}$$

this would not change the rank of the tree, since the degree of the added formula would be zero. When we form (**) we are actually doing something that will increase the ranks of all the end points of such a tree to at most 2^{μ_2}. By the rank Lemma 2 this will raise the rank of Y_n to at most

$$2^{\mu_2} + 2^{\mu_1} \leq 2^{\max(\mu_1,\mu_2)} + 2^{\max(\mu_1,\mu_2)} = 2^{\max(\mu_1,\mu_2)+1} = 2^{\mu}.$$

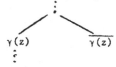

Theorem 6. Every closed cut-bounded tableau of degree $\leq n$ and rank μ can be transformed into a cut-free closed tableau with the same origin branch and with a rank $\leq 2_{n+1}(\mu)$.

Proof. We use complete induction on the degree n. If $n = 0$ we have by proposition 21 a closed cut-free tableau of rank $\leq 2^{\mu} = 2_1(\mu)$ with the same origin branch. Now let $n \neq 0$ and let the proposition hold for all closed cut-bounded trees of degree $< n$. By proposition 22 a given closed cut-bounded tableau of degree n and rank $\leq \mu$ can be transformed into a closed cut-bounded tree of degree $\leq n-1$ and rank $\leq 2^{\mu}$. By the induction hypothesis this tree can be transformed into a cut-free closed tableau with the same origin branch and with a rank

$\leq 2_{(n-1)+1}(2^{\mu}) = 2_{n+1}(\mu)$.

Thus a tableau proof in Ω for a formula P that contains cuts of degree $\leq n$ and has a rank μ can be transformed into a cut-free tableau proof in Ω of the same formula that has a rank β such that

1) $\beta \leq 2_{n+1}(\mu)$

2) if μ is less than some ϵ-number ϵ, then $\beta < \epsilon$.

It follows that:

Theorem 7. There is no formula P such that both P and $\neg P$ are provable in Ω with cut-bounded proofs.

For again let

$$\begin{array}{c} F\ P \\ \mathcal{T}_1 \end{array}$$

be a cut-bounded proof of P and

$$\begin{array}{c} F\neg P \\ \mathcal{T}_2 \end{array}$$

be a cut-bounded proof of $\neg P$.

Then

$$\dfrac{\begin{array}{c} T\neg P \\ FP \\ \mathcal{T}_1 \end{array} \qquad \begin{array}{c} F\neg P \\ \mathcal{T}_2 \\ \end{array}}{F0 = 0'}\ \text{CUT}$$

is a normal proof of $0 = 0'$. By theorem 6 one could obtain a cut-free proof of $0 = 0'$ from this proof. But this is impossible since $F0 = 0'$ does not close a branch and no rules are applicable to $F0 = 0'$. This yields the consistency of Ψ, as was discussed at the end of the last section.

We can also now see

<u>Theorem 8</u>. Any universal closure of a formula provable in Ψ has a cut-free proof in Ω of rank $< \epsilon_0$.

For proposition 20 tells us that any closed formula provable in Ψ has a cut-bounded proof in Ω of rank $< \omega + \omega$. If n is the degree of the resulting proof in Ω, we can transform it to a cut-free proof of rank $< 2_{n+1}(\omega+\omega) < \epsilon_0$.

§5. The Incompleteness of Ψ (Missing Provable Ordinals)

We now assume the rank relation and ordinal functions mentioned in the introduction occur in Ψ and Ω. We remark again that since we are mentioning only ordinal functions explicitly, we have dropped the circles in the notations, e.g. $+$ and $<$ denote $\oplus$ and $\circledless$ unless mention is made to the contrary.

We can express the <u>$<$-progressiveness</u> of a formula $A(t)$ by the formula

$$P_y A(y) \equiv \forall y((\forall x)(x < y \supset A(x)) \supset A(y))$$

<u>Transfinite induction</u> over the ordinals $< t$ is expressed by the formula

$$J_x(A(x),t) \equiv P_y A(y) \supset (\forall x)(x < t \supset A(x)).$$

If $J_x(A(x),t)$ is provable for every formula $A(x)$, we say t is a <u>provable ordinal</u>.

In this paragraph we will investigate to what extent this formalized transfinite induction can be derived in the systems Ψ and Ω and we will use the formalized transfinite induction as a criterion for the incompleteness of the system Ψ.

We will be giving many proofs in Ψ, using the derivable rules freely. In this section we collect together some arithmetically true signed formulas that will be used with the <u>axiom</u> derived rule and whose conjugates will be used to close branches, and we will also give proofs of some basic formulas that will be used in later proofs with the theorem rule and derived closure condition d.

Since $z < 0$ is a numerically false formula for every numeral z,

(A1) $F \ a < 0$

is arithmetically true. By the transitivity of the <-relation,

(A2) $T(c < b \wedge b < a) \supset c < a$

is arithmetically true. Since z^* is the <-successor of z, one has the arithmetically true formulas

(A3) $T \ a < a^*$,

(A4) $T \ c < a^* \supset (c < a \vee c = a)$.

With the properties of $+$ and $E(z)$ we obtain the arithmetically true formulas

(A5) $T \ a^* = a + E(0)$,

(A6) $T \ 0 + E(c) = E(c)$,

(A7) $T \ (b + E(f(a,b))) + E(f(a,b)) = b + E(f(a,b)^*)$

for every two-place function f.

We note that

$$\mu(\nu(0) = \mu(0) = 0,$$

$$\mu(\nu(z^*)) = \mu(\nu(z)') = \mu(\nu(z))^*.$$

From this one sees that $\mu(\nu(z)) = z$ for all $z < \omega$. One thus has the arithmetically true formulas

(A8) $T \ \mu(0) = 0$,

(A9) $T \ \mu(a)^* = \mu(a')$,

(A10) $T \ a < \omega \supset \mu(\nu(a)) = a$.

From 8 in the introduction:

(A11) $T \ a < b + E(\Psi(a,b)^*)$,

(A12) $T \ a < b + E(c) \supset (\Psi(a,b) < c \vee c = 0)$.

In the following we consider a fixed formula $A(t)$. We use the abbreviations:

$$A^*(t) \equiv (\forall x)(x < t \supset A(x))$$

$$A_1(t) \equiv (\forall y)(A^*(y) \supset A^*(y + E(t)))$$

$$A_1^*(t) \equiv (\forall x_1)(x_1 < t \supset A_1(x_1))$$

$$P \equiv (\forall y)(A^*(y) \supset A(y))$$

$$P_1 \equiv (\forall y_1)(A_1^*(y_1) \supset A_1(y_1))$$

The formulas P and P_1 express the $<$-progressiveness of $A(t)$ and $A_1(t)$ respectively with respect to the argument place of t. The corresponding transfinite induction over numbers $< t$ is expressed by

$$J_x(A(x),t) \equiv P \supset A^*(t),$$

$$J_x(A_1(x),t) \equiv P_1 \supset A_1^*(t).$$

By proposition 2, all formulas of the form $(\forall x)B(x) \supset B(t)$ are provable, since for any $B(t)$ we have the partial tableau

F $(\forall x)B(x) \supset B(t)$

T $(\forall x)B(x)$

F $B(t)$

T $B(t)$

which can be extended to a closed tableau.

As a result of this we see that the following formulas are provable:

(T1) $A^*(a) \supset (c < a \supset A(c))$,

(T2) $A^*(a^*) \supset (a < a^* \supset A(a))$,

(T3) $A^*(b + E(\Psi(a,b)^*)) \supset (a < b + E(\Psi(a,b)^*) \supset A(a))$,

(T4) $A_1(c) \supset (A*(0) \supset A*(0 + E(c)))$,

(T5) $A_1(\Psi(a,b)) \supset (A*(b) \supset A*(b + E(\Psi(a,b))))$,

(T6) $A_1(\Psi(a,b)) \supset (A*(b + E(\Psi(a,b))) \supset A*((b + E(\Psi(a,b))) + E(\Psi(a,b))))$,

(T7) $A_1*(c) \supset (\Psi(a,b) < c \supset A_1(\Psi(a,b))$,

(T8) $P \supset (A*(a) \supset A(a))$.

Induction Proofs in Ψ.

Proposition 23. The following formulas are derivable in Ψ:

I. $J_x(A(x),0)$,

II. $J_x(A(x),a) \supset (b < a \supset J_x(A(x),b))$,

III. $J_x(A(x),a) \supset J_x(A(x),a*)$,

IV. $J_x(A(x),\omega)$,

V. $J_x(A_1(x),c) \supset J_x(A(x),E(c))$.

In giving these proofs we will use the following conventions. The proofs will be given as tableaux in which the derived rules are allowed to be used. To aid the reader the formulas in the proof will be numbered[9] and the numbers of the premises of an inference will be indicated next to the line(s) to the conclusion(s) of the inference. The rule applied will also be indicated, with the use of the following abbreviations used: CI, AX, REP, EQ, and TH will mean respectively complete induction (in any of its forms - cf. propositions 8 and 9), axiom, replacement, equality, and theorem. When a step consists only in making or writing out an abbreviation, AB will appear. The primitive closure of a branch will be denoted by $\#_a$, $\#_b$, and $\#_c$ respec-

[9] The number assigned to a formula will be found near the left-hand margin of the line on which the formula occurs in the text.

tively, while the fact that a branch can be extended to a closed branch because it satisfies conditions a, b, c, and d of the corollary to proposition 3 will be indicated by $\$_a$, $\$_b$, $\$_c$, or $\$_d$ respectively. The formulas needed in any rule or closure, if any, will always be indicated by their numbers, axiom number, or theorem number. When a tableau must be continued from one page to another, the open branches will be numbered with corresponding circled numbers on both pages.

The proof of formula I will be explained to make the notation clear.

Proof of formula I.

1 $F\ J_x(A(x),0)$

| AB;1

2 $F\ P \supset (\forall x)(x < 0 \supset A(x))$

| α;2

3 $F(\forall x)(x < 0 \supset A(x))$

| δ;3

4 $F\ a < 0 \supset A(a)$

| α;4

5 $T\ a < 0$

$\#_a$;5 cf.(A1)

Thus formula 2 is obtained by writing out the definition of $J_x(A(x),0)$, which is formula 1. Formula 3 is the result of applying an α rule to formula 2 (to get α_2). Formula 4 is the result of applying a δ rule to the formula 3. And formula 5 is the result of applying an α rule to the formula 4. Here formula 5 causes the tableau to close, i.e., $T\ a < 0$ is an arithmetically unsatisfiable signed formula, as we have noted above with axiom (A1).

Proof of formula II.

1 $F\ J_x(A(x),a) \supset (b < a \supset J_x(A(x),b))$

AB;1

2 $F\ (P \supset (\forall x)(x < a \supset A(x))) \supset ((b < a) \supset (P \supset (\forall x)(x < b \supset A(x))))$

α;2

3 $T\ P \supset (\forall x)(x < a \supset A(x))$

α;2

4 $F\ b < a \supset (P \supset (\forall x)(x < b \supset A(x)))$

α;4

5 $T\ b < a$

α;4

6 $F\ P \supset (\forall x)(x < b \supset A(x))$

α;6

7 $T\ P$

α;6

8 $F\ (\forall x)(x < b \supset A(x))$

δ;8

9 $F\ c < b \supset A(c)$

β;3

10 FP

$\$_b$;7;10

11 $T\ (\forall x)(x < a \supset A(x))$

(1)

(1)

γ;11

12 $T\ c < a \supset A(c)$

α;9

13 $T\ c < b$

α;9

14 $F\ A(c)$

β;12

15 $T\ A(c)$

$\$_b$;14,15

16 $F\ c < a$

$\#_a$;5,13,16 (cf. (A2))

Proof of formula III.

1 $F\ J_x(A(x),\ a) \supset J_x(A(x), a^*)$

AB;1

2 $F\ (P \supset (\forall x)(x < a \supset A(x))) \supset (P \supset (\forall x)(x < a^* \supset A(x)))$

α;2

3 $T\ P \supset (\forall x)(x < a \supset A(x))$

α;2

4 $F\ P \supset (\forall x)(x < a^* \supset A(x))$

α;4

5 $T\ P$

AB;5

(2)

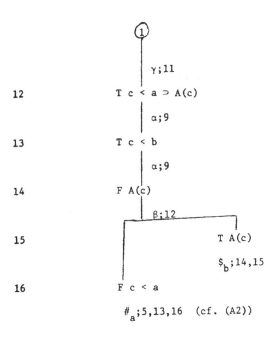

(2)

6 $T(\forall y)((\forall x)(x < y \supset A(x)) \supset A(y))$

$\alpha;4$

7 $F(\forall x)(x < a^* \supset A(x))$

$\beta;3$

8 FP

$\$_b;5,8$

9 $T(\forall x)(x < a \supset A(x))$

$\delta;7$

10 $F\ c < a^* \supset A(c)$

$\alpha;10$

11 $T\ c < a^*$

$\alpha;10$

12 $F\ A(c)$

$\gamma;9$

13 $T\ c < a \supset A(c)$

$\beta;13$

14 $T\ A(c)$

$\$_b;12,14$

15 $F\ c < a$

$\gamma;6$

16 $T\ (\forall x)(x < a \supset A(x)) \supset A(a)$

$\beta;16$

17 $F(\forall x)(x < a \supset A(x))$

$\$_b;9,17$

(3)

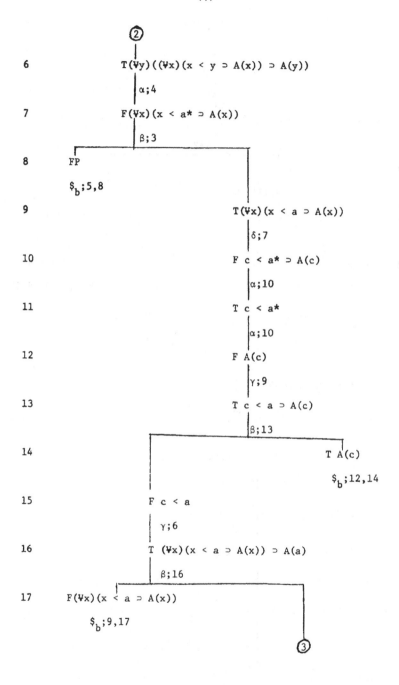

(3)

18 $T\ A(a)$

$AX;(A4)$

19 $T\ c < a^{*} \supset (c < a \vee c = a)$

$\beta;19$

20 $F\ c < a^{*}$

$\#_a;11,20$

21 $T\ c < a \vee c = a$

$\beta;21$

22 $T\ c = a$

$\$_c;12,18,22$

23 $T\ c < a$

$\#_a;15,23$

Proof of formula IV.

1 $F\ J_x(A(x),\omega)$

$AB;1$

2 $F\ P \supset (\forall x)(x < \omega \supset A(x))$

$\alpha;2$

3 $T\ P$

$\alpha;2$

4 $F(\forall x)(x < \omega \supset A(x))$

$\delta;4$

5 $F\ a < \omega \supset A(a)$

$\alpha;5$

(5)

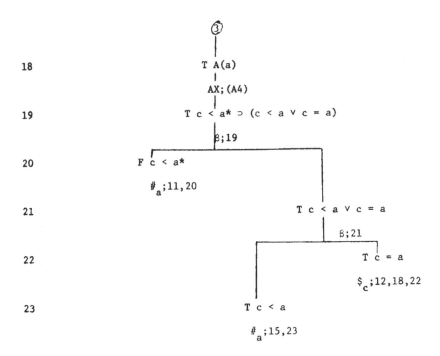

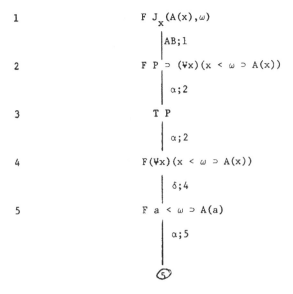

(5)

6 T $a < \omega$

α;5

7 F $A(a)$

CUT

8 T$(\forall x)(x < a^* \supset A(x))$

γ;8

9 T $a < a^* \supset A(a)$

β;9

10 F $a < a^*$

$\#_a$;10

11 T $A(a)$

$\$_b$;7,11

12 F$(\forall x)(x < a^* \supset A(x))$

TH;III

13 T $J_x(A(x), a) \supset J_x(A(x), a^*)$

β;13

14 T $J_x(A(x), a^*)$

AB;14

15 T $P \supset (\forall x)(x < a^* \supset A(x))$

β;15

16 F P

$\$_b$;3,16

17 T$(\forall x)(x < a^* \supset A(x))$

$\$_b$;12,17

18 F $J_x(A(x), a)$

AX;(A10)

(6)

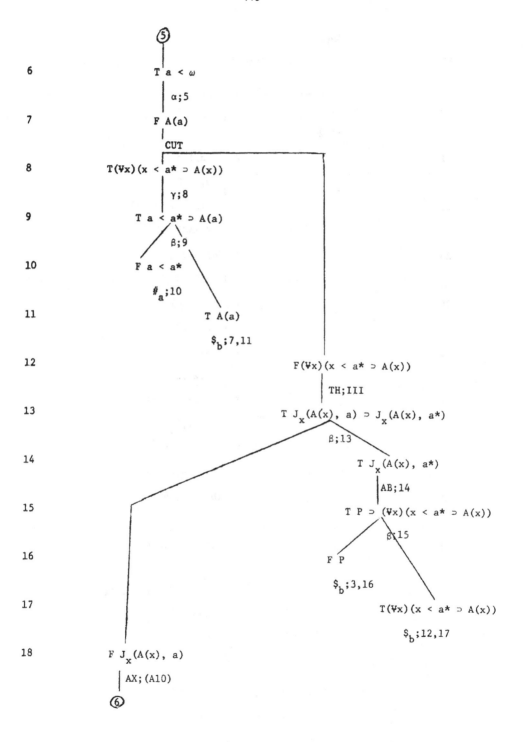

(6)

19 $T\ a < \omega \supset \mu(\nu(a)) = a$

β;19

20 $F\ a < \omega$

$\#_a$;6,20

21 $T\ \mu(\nu(a)) = a$

EQ;18,21

22 $F\ J_x(A(x),\ \mu(\nu(a)))$

AX;(A8)

23 $T\ \mu(0) = 0$

TH;I

24 $T\ J_x(A(x),0)$

EQ;23,24

25 $T\ J_x(A(x),\ \mu(0))$

CI;22,25

26 $T\ J_x(A(x),\ \mu(b))$

CI;22,25

27 $F\ J_x(A(x),\ \mu(b'))$

AX;(A9)

28 $T\ \mu(b)^* = \mu(b')$

EQ;27,28

29 $F\ J_x(A(x),\ \mu(b)^*)$

TH;III

30 $T\ J_x(A(x),\ \mu(b)) \supset J_x(A(x),\ \mu(b)^*)$

β;30

(7)

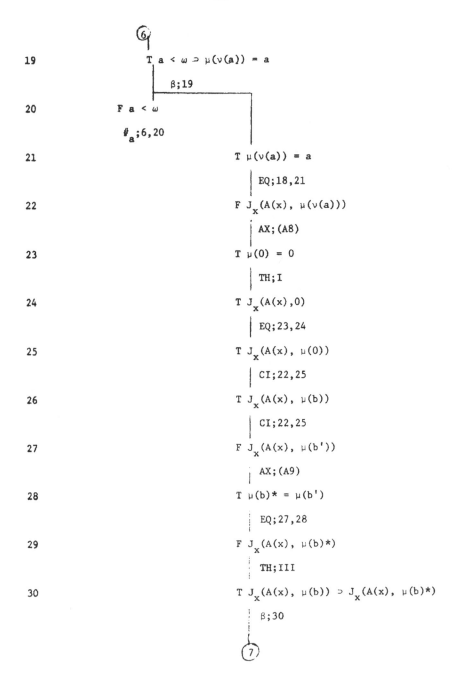

(7)

31 $F\ J_x(A(x),\ \mu(b))$

$\$_b$;26,31

32 $T\ J_x(A(x),\ \mu(b)^*)$

$\$_b$;29,32

We will break the proof of formula V up into three parts:

Lemma 1. The set $\{T\ c = 0,\ T\ P,\ F\ A_1(c)\}$ has a closed tableau.

Proof. Such a tableau is the following:

1 $T\ c = 0$

2 $T\ P$

3 $F\ A_1(c)$

EQ;1,3

4 $F\ A_1(0)$

AB;4

5 $F(\forall y)(A^*(y) \supset A^*(y + E(0)))$

δ;5

6 $FA^*(a) \supset A^*(a + E(0))$

AX;(A5)

7 $T\ a^* = a + E(0)$

(7)

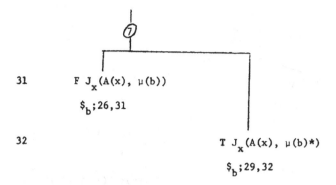

(7)

CUT

8 T $A^*(a) \supset A^*(a^*)$

$\$_c$;6,7,8

9 F $A^*(a) \supset A^*(a^*)$

AB;2

10 T $(\forall y)(A^*(y) \supset A(y))$

α;9

11 T $A^*(a)$

AB;11

12 T $(\forall x)(x < a \supset A(x))$

α;9

13 F $A^*(a^*)$

AB;13

14 F $(\forall x)(x < a^* \supset A(x))$

δ;14

15 F $e < a^* \supset A(e)$

γ;12

16 T $e < a \supset A(e)$

α;15

17 T $e < a^*$

α;15

18 F $A(e)$

AX;(A4)

19 T $e < a^* \supset (e < a \vee e = a)$

β;19

(8)

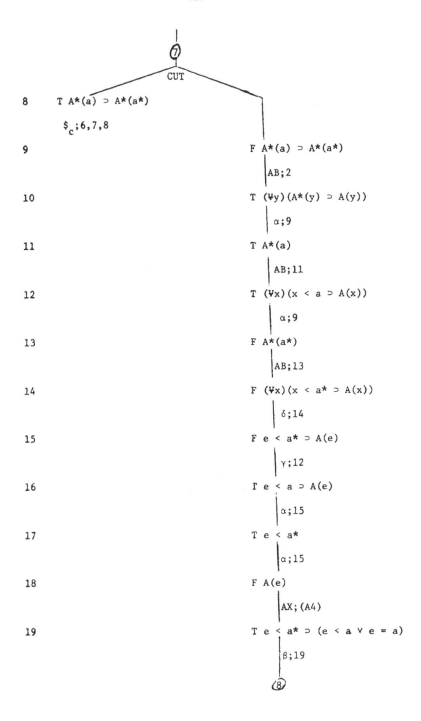

(8)

20 F e < a*

$\$_b$;17,20

21 T e < a ∨ e = a

β;20

22 T e < a

β;16

23 F e < a

$\#_a$;22,23

24 T A(e)

$\$_b$;18,24

25 T e = a

γ;10

26 T A*(a) ⊃ A(a)

β;26

27 F A*(a)

$\$_b$;11,27

28 T A(a)

$\$_c$;18,25,28

Let us call the tree starting at line 4 in the above tableau

$$\tilde{\mathcal{T}}_1^c$$

(for an arbitrary parameter c).

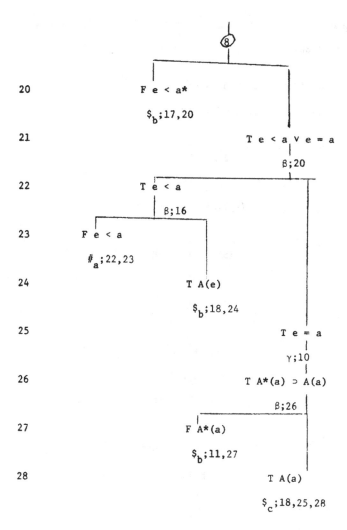

<u>Lemma 2</u>. The set $\{F\ c = 0,\ T\ A_1{*}(c),\ F\ A_1(c)\}$ has a closed tableau.

Proof. Such a tableau is the following:

1 $F\ c = 0$

2 $T\ A_1{*}(c)$

3 $F\ A_1(c)$

AB; 3

4 $F\ (\forall y)(A{*}(y) \supset A{*}(y + E(c)))$

δ; 4

5 $F\ A{*}(b) \supset A{*}(b + E(c))$

α; 5

6 $T\ A{*}(b)$

α; 5

7 $F\ A{*}(b + E(c))$

AB; 7

8 $F\ (\forall x)(x < b + E(c) \supset A(x))$

δ; 8

9 $F\ d < b + E(c) \supset A(d)$

α; 9

10 $T\ d < b + E(c)$

α; 9

11 $F\ A(d)$

AX; (A12)

⑨

(9)

12 $T\ d < b + E(c) \supset (\Psi(d,b) < c \lor c = 0)$

$\beta;12$

13 $F\ d < b + E(c)$

$\#_a;10,13$

14 $T\ \Psi(d,b) < c \lor c = 0$

$\beta;14$

15 $T\ c = 0$

$\#_a;1,15$

16 $T\ \Psi(d,b) < c$

TH; (T7)

17 $T\ A_1{*}(c) \supset (\Psi(d,b) < c \supset A_1(\Psi(d,b)))$

$\beta;17$

18 $F\ A_1{*}(c)$

$\$_b;2,18$

19 $T\ \Psi(d,b) < c \supset A_1(\Psi(d,b))$

$\beta;19$

20 $F\ \Psi(d,b) < c$

$\#_a;16,20$

21 $T\ A_1(\Psi(d,b))$

AX; (A11)

22 $T\ d < b + E(\Psi(d,b){*})$

TH; (T3)

23 $T\ A{*}(b + E(\Psi(d,b){*})) \supset (d < b + E(\Psi(d,b){*}) \supset A(d))$

$\beta;23$

(10)

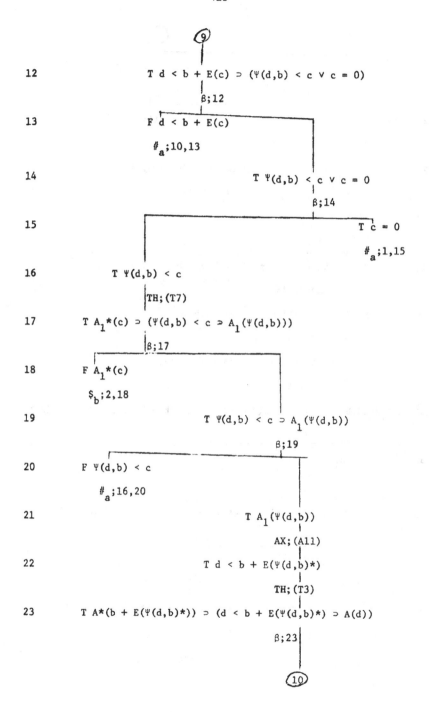

(10)

24 $T\ d < b + E(\Psi(d,b)^*) \supset A(d)$

$\beta;24$

25 $T\ A(d)$

$\$_b;11,24$

26 $F\ d < b + E(\Psi(d,b)^*)$

$\#_a;22,26$

27 $F\ A^*(b + E(\Psi(d,b)^*))$

AX:(A7)

28 $T\ (b + E(\Psi(d,b))) + E(\Psi(d,b)) = b + E(\Psi(d,b)^*)$

EQ;27,28

29 $F\ A^*((b + E(\Psi(d,b))) + E(\Psi(d,b)))$

TH;(T5)

30 $T\ A_1(\Psi(d,b)) \supset (A^*(b) \supset A^*(b + E(\Psi(d,b))))$

TH;(T6)

31 $T\ A_1(\Psi(d,b)) \supset (A^*(b + E(\Psi(d,b))) \supset A^*((b + E(\Psi(d,b))) + E(\Psi(d,b))))$

$\beta;31$

32 $F\ A_1(\Psi(d,b))$

$\$_b;21,32$

33 $T\ A^*(b + E(\Psi(d,b))) \supset A^*((b + E(\Psi(d,b))) + E(\Psi(d,b)))$

$\beta;30$

34 $F\ A_1(\Psi(d,b))$

$\$_b;21,34$

35 $T\ A^*(b) \supset A^*(b + E(\Psi(d,b)))$

$\beta;35$

(11)

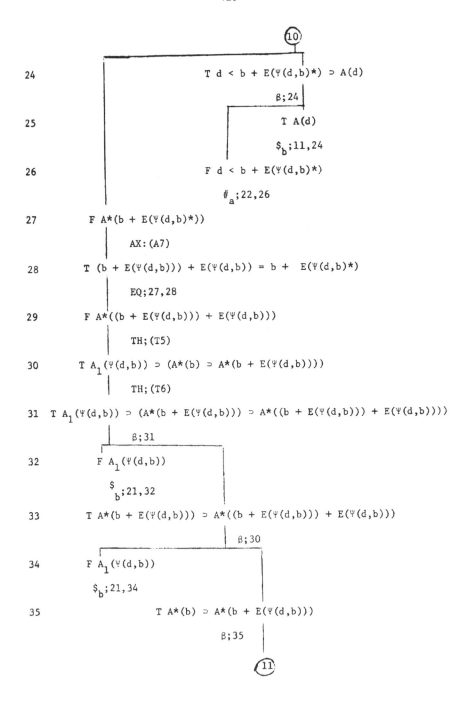

(11)

36 F A*(b)

$\$_b$;6,36

37 T A*(b + E(Ψ(d,b)))

β;33

38 F A*(b + E(Ψ(d,b)))

$\$_b$;37,38

39 T A*((b + E(Ψ(d,b))) + E(Ψ(d,b)))

$\$_b$;29,39

Let us call the tree starting at line 4 in the above tableau

$$\mathcal{T}_2^c$$

(for an arbitrary parameter c).

<u>Lemma 3</u>. The set $\{T\ A_1(c),\ F\ A^*(E(c))\}$ has a closed tableau.

Proof. Such a tableau is the following:

1 T A_1(c)

2 F A*(E(c))

AX; (A6)

3 T 0 + E(c) = E(c)

CUT

4 F A*(0)

AB;4

(12) (13)

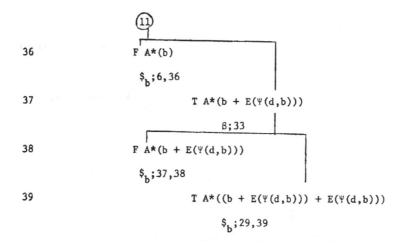

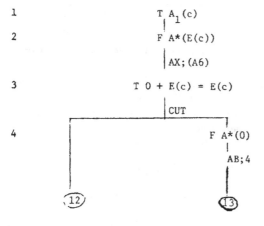

(12) | (13)

5 $F(\forall x)(x < 0 \supset A(x))$ — δ;5

6 $F\ d < 0 \supset A(d)$ — α;6

7 $T\ d < 0$

$\#_a$;7

8 $T\ A^*(0)$ — TH;(T4)

9 $T\ A_1(c) \supset (A^*(0) \supset A^*(0 + E(c)))$ — β;9

10 $F\ A_1(c)$

$\$_b$;1,10

11 $T\ A^*(0) \supset A^*(0 + E(c))$ — β;11

12 $F\ A^*(0)$

$\$_b$;8,12

13 $T\ A^*(0 + E(c))$

$\$_c$;2,3,13

Let us call the tree starting with line 3 of the above tableau

$$\mathcal{T}_3^c$$

(for an arbitrary parameter c).

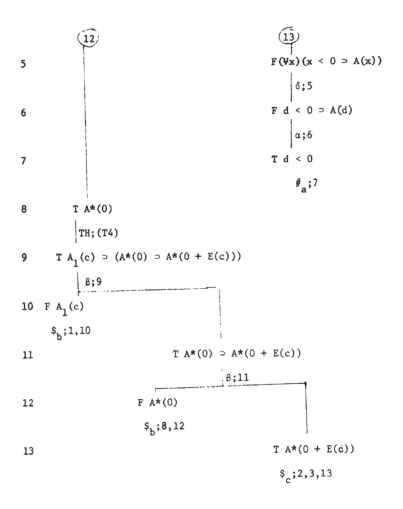

Proof of formula V.

1 $F\ J_x(A_1(x,g) \supset J_x(A(x),E(g))$

AB;1

2 $F(P_1 \supset A_1^*(g)) \supset (P \supset A^*\ (E(g)))$

α;2

3 $T\ P_1 \supset A_1^*(g)$

α;2

4 $F\ P \supset A^*(E(g)))$

α;4

5 $T\ P$

α;4

6 $F\ A^*(E(g))$

β;3

7 $T\ A_1^*(g)$

CUT

8 $F\ A_1(g)$

CUT

9 $F\ g = 0$

γ_2^g

10 $T\ g = 0$

γ_1^g

11 $T\ A_1(g)$

γ_3^g

(14)

(14)

12 F P_1

AB;12

13 $F(\forall y_1)(A_1^*(y_1) \supset A_1(y_1))$

δ;13

14 $F\ A_1^*(c) \supset A_1(c)$

α;14

15 $T\ A_1^*(c)$

α;14

16 $F\ A_1(c)$

CUT

17 $T\ c = 0$

$\mathcal{T}_1^c$

18 $F\ c = 0$

$\mathcal{T}_2^c$

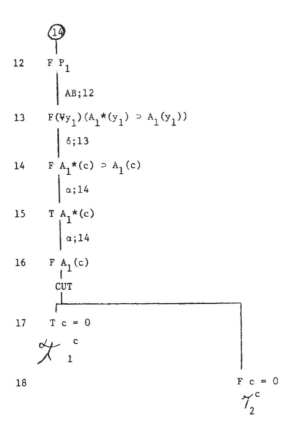

Proposition 24. If $J_x(A(x),z)$ is derivable in Ψ for some numeral z,

a) $J_x(A(x),z_1)$ for all numerals $z_1 < z$, and

b) $J_x(A(x),z^*)$

are also derivable in Ψ.

Proof. It is assumed that there is a closed tableau $\mathcal{T}$ for $J_x(A(x),z)$. The proofs of the formulas in question are then as follows:

a)

1 $F\ J_x(A(x), z_1)$

TH; II

2 $T\ J_x(A(x), z) \supset (z_1 < z \supset J_x(A(x), z_1)$

β; 2

3 $F\ J_x(A(x), z)$

4 $T\ z_1 < z \supset J_x(A(x), z_1)$

β; 4

5 $F\ z_1 < z$

$\#_a$; 5

6 $T\ J_x(A(x), z_1)$

$\$_b$; 1, 6

b)

1 $F\ J_x(A(x), z^*)$

TH; III

2 $T\ J_x(A(x), z) \supset J_x(A(x), z^*)$

β; 2

3 $F\ J_x(A(x), z)$

4 $T\ J_x(A(x), z^*)$

$\$_b$; 1, 4

We define the <u>iterated exponential function</u> $E_n(\alpha)$ recursively for natural numbers n:

$$E_0(\alpha) = \alpha$$

$$E_{n+1}(\alpha) = E(E_n(\alpha)).$$

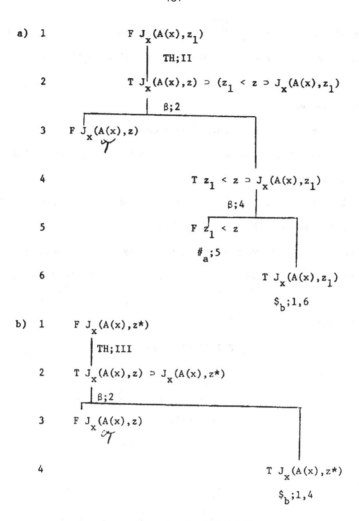

Proposition 25. If $J_x(A(x),z)$ is derivable in Ψ for a fixed numeral z and for every formula $A(t)$, then $J_x(A(x),E_n(z))$ is also derivable in Ψ for every formula $A(t)$ and every natural number n.

The proof is by complete induction on n.

a) For $n = 0$ one obtains the assertion with the replacement rule since $E_0(z) = z$.

b) Let n be > 0, and let the assertion hold for $n - 1$. Let $A(t)$ be an arbitrary formula and let $J_x(F(x),z)$ be derivable in Ψ for every formula $F(t)$. By the induction hypothesis, in this case $J_x(A_1(x),E_{n-1}(z))$ is derivable in Ψ, say with a tableau $\mathcal{T}$. We then form a proof for $J_x(A(x),E_n(z))$ as follows:

1 $\quad F\ J_x(A(x),E_n(z))$

$\quad$ | AX

2 $\quad T\ E_n(z) = E(E_{n-1}(z))$

$\quad$ | EQ;1,2

3 $\quad F\ J_x(A(x),E(E_{n-1}(z))$

$\quad$ | TH;V

4 $\quad T\ J_x(A_1(x),E_{n-1}(z)) \supset J_x(A(x),E(E_{n-1}(z))$

$\quad$ | β;4

	Left branch	Right branch
5	$F\ J_x(A_1(x),E_{n-1}(z))$	
6	$\mathcal{T}$	$T\ J_x(A(x),E(E_{n-1}(z))$
		$\$_b$;3,6

<u>Theorem 9</u>. For every numeral $z < \epsilon_0$, $J_x(A(x),z)$ is derivable in Ψ.

Proof. If $z < \epsilon_0$, by statement 11 and 12 of the introduction (recall $E(a) = 2^a$), there is a natural number n with $z < E_n(\omega^*)$. By IV, $J_x(A(x), \omega)$ is derivable in Ψ for every formula $A(t)$. Then, by proposition 24b $J_x(A(x), \omega^*)$ is derivable; by Proposition 25 $J_x(A(x),E_n(\omega^*)$ is derivable, and by Proposition 24a, $J_x(A(x),z)$ is derivable.

<u>Definition of the system</u> Ψ_α. The system Ψ_α (Ψ_α') differs from the system Ψ (Ψ') only in the fact that the formulas

$$F\ J_x(F(x),\alpha)$$

are allowed to be used to close a branch (for arbitrary $F(t)$). Ψ_α and Ψ_α' can be seen to be equivalent just as in the case of Ψ and Ψ'.

Propositions 24 and 25 clearly hold for the system Ψ_α also.

<u>Theorem 10</u>. If $\varepsilon_\nu \leq \alpha < \varepsilon_{\nu+1}$, then for every numeral $z < \varepsilon_{\nu+1}$, $J_x(A(x)$ is derivable in Ψ_α.

Proof. By statements 11 and 12 of the introduction, there is a natural number n with $z < E_n(\varepsilon_\nu{*})$, since $z < \varepsilon_{\nu+1}$. It is clear that $J_x(A(x),\alpha)$ is provable in Ψ_α. Since $\varepsilon_\nu \leq \alpha$, by Proposition 24a, $J_x(A(x),\varepsilon_\nu)$ is also provable; by proposition 24b $J_x(A(x),\varepsilon_\nu^*)$ is provable; and by proposition 25 $J_x(A(x),E_n(\varepsilon_\nu{*}))$ is provable. With Proposition 24a it then follows that $J_x(A(x),z)$ is also provable in Ψ_α.

Induction derivations in Ω. We now look for cut-free proofs in the system Ω for the formula $J_x(A(x),z)$. A limit depending on the numeral z will be given for the rank of such a derivation. For this let $A(t)$ be an arbitrary formula of the system Ω. In this section when we say that a formula in Ω is derivable with rank α, this will always mean that there is a cut-free derivation in Ω (without derived rules and closure conditions) that is of rank $\leq \alpha$.

Lemma 1. For every numeral $z < \epsilon_0$, $J_x(A(x),z)$ is derivable in Ω with a rank $< \epsilon_0$.

Proof. By Theorem 9, for $z < \epsilon_0$ the formula $J_x(A(x),z)$ is derivable in Ψ. Then by Theorem 8, the formula has a cut-free derivation in Ω with a rank $< \epsilon_0$.

Lemma 2. If for every formula $A(t)$, $J_x(A(x), \epsilon_\nu)$ is derivable in Ω with a rank $\leq \epsilon_\nu + 1$ and $z < \epsilon_{\nu+1}$, $J_x(A(x),z)$ is derivable in Ω with a rank $< \epsilon_{\nu+1}$.

Proof. It is easy to see that the formula $J_x(A(x),z)$, which has a proof in Ψ_{ϵ_ν} by Theorem 10, has a derivation in Ψ' if formulas $F\ J_x(F(x), \epsilon_\nu)$ are also allowed to close branches. If we look at the proof that shows that anything provable in Ψ' is provable in Ω, we see that we first transformed the tableau so that rules of Ω were used in it (Lemma 1, §3).

In doing this we do not raise the rank of the resulting tree above its original rank, say the integer m. We then extended branches that were not closed according to the closure conditions of Ω (Lemma 2, §3). To do this for branches closing by conditions of Ψ' required only finite extensions. In the present case we may have to make extensions for branches closing because of formulas $F\, J_x(F(x),\epsilon_\nu)$. But by the induction hypothesis of the lemma, these formulas have proofs of rank $\leq \epsilon_\nu + 1$. Thus all the end points of the branches of the tree are raised to at most $\epsilon_\nu + 1$, and so by the rank Lemma 2, the rank of the resulting proof in Ω is $\leq \epsilon_\nu + 1 + m$. Now this proof may still have cuts. So we apply Theorem 5 to obtain a proof of rank $\leq E_{n+1}(\epsilon_\nu + 1 + m) < \epsilon_{\nu+1}$. One thus obtains a primitive derivation in Ω of rank $< \epsilon_{\nu+1}$ for $J_x(A(x),z))$.

Lemma 3. If ϵ is an ϵ-number and $J_x(A(x),z)$ is derivable in Ω with a rank $\mu < \epsilon$ for every numeral $z < \epsilon$, then $J_x(A(x),\epsilon)$ is derivable in Ω with a rank $\leq \epsilon + 1$.

Proof. When $z < \epsilon$, also $z^* < \epsilon$. Then by the assumption there is a cut-free derivation $\mathcal{T}_{z^*}$ for $J_x(A(x),z^*)$ in Ω. With a rank $\mu_{z^*} < \epsilon$. With these derivations we form the following proof (with cuts) of $J_x(A(x),\epsilon)$.

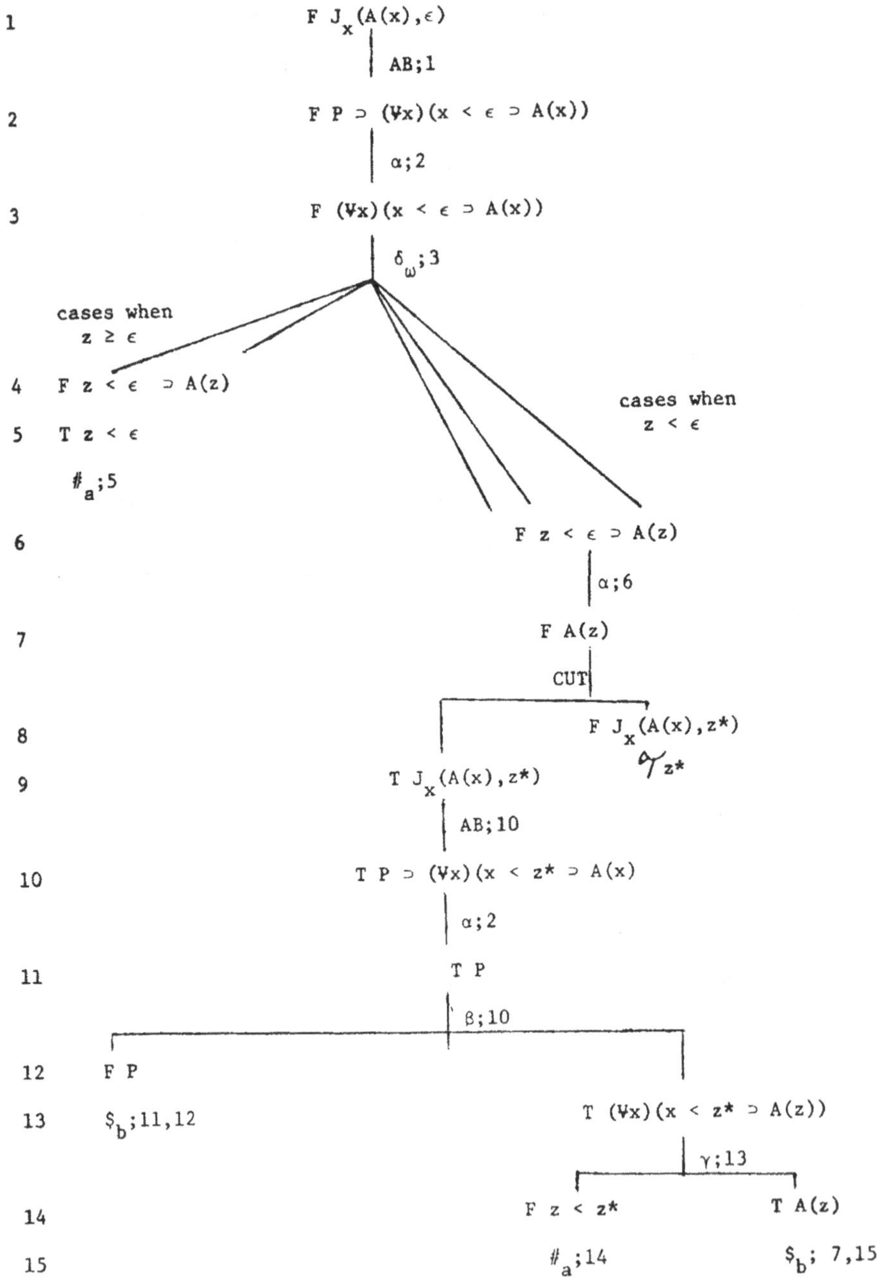
1 F J_x(A(x),ϵ)
AB;1
2 F P ⊃ (∀x)(x < ϵ ⊃ A(x))
α;2
3 F (∀x)(x < ϵ ⊃ A(x))
δ_ω;3
cases when z ≥ ϵ
4 F z < ϵ ⊃ A(z)
5 T z < ϵ
#_a;5
cases when z < ϵ
6 F z < ϵ ⊃ A(z)
α;6
7 F A(z)
CUT
8 F J_x(A(x),z*)
9 T J_x(A(x),z*)
AB;10
10 T P ⊃ (∀x)(x < z* ⊃ A(x)
α;2
11 T P
β;10
12 F P
13 $_b;11,12
T (∀x)(x < z* ⊃ A(z))
γ;13
14 F z < z*
T A(z)
15 #_a;14
$_b; 7,15

We must now consider the rank of this proof, or rather of the proof that results when the single cut is eliminated. We notice that in the proof given the rank of formula 7 is $\leq \max\{\omega, \mu_{z*}\} + 1 < \epsilon$. By our cut-elimination proposition we can eliminate this cut without raising the rank of formula 7 above ϵ. So the ranks of all of the successors of formula 3 have ranks $< \epsilon$. Thus formula 3 will have rank $\leq \epsilon$, and formula 2, which is the formula we are proving, will have rank $\leq \epsilon + 1$.

The general proposition on the ranks of induction derivations in Ω is:

Theorem 11. For every ϵ-number ϵ:

a) For $z < \epsilon$, $J_x(A(x),z)$ is derivable in Ω with a rank $< \epsilon$.

b) $J_x(A(x),\epsilon)$ is derivable in Ω with a rank $\leq \epsilon + 1$.

The proof is by transfinite induction on ϵ.

I. If $\epsilon = \epsilon_0$, we have a) by Lemma 1, and then b) follows from a) with Lemma 3.

II. Let $\epsilon_0 < \epsilon$ and let the theorem hold for all ϵ-numbers $< \epsilon$. We consider an arbitrary numeral $z < \epsilon$. If $z < \epsilon_0$ we have assertion a) by I. Otherwise by statement 11 of the Introduction, there is a largest ϵ-number $\epsilon_\nu \leq z$. Since $z < \epsilon$, it must be true that $z < \epsilon_{\nu+1} \leq \epsilon$. By the induction hypothesis $J_x(F(x), \epsilon_\nu)$ is derivable in Ω with a rank $\leq \epsilon_\nu + 1$ for every formula $F(t)$. Then by Lemma 2 $J_x(A(x),z)$ is derivable in Ω with a rank $< \epsilon_{\nu+1} \leq \epsilon$. This proves assertion a). Assertion b) then follows by Lemma 3.

One sees from Theorem 11 that the induction formula $J_x(A(x),z)$ is derivable in Ω for <u>every</u> numeral z. It will be shown that this derivability does not occur in Ψ or in any system Ψ_α.

<u>Limits on the derivability of transfinite induction</u>. We now wish to consider which numerals z are such that an induction formula $J_x(A(x),a)$ is not derivable in Ψ or in a system Ψ_α. To decide this we need a lower bound depending on z for the rank of a cut-free derivation of $J_x(A(x),z)$ in Ω. To find such a lower bound we consider instead of Ω an extended system $\bar{\Omega}$.

<u>The system $\bar{\Omega}$</u>. The primitive symbols of the system $\bar{\Omega}$ consist of the primitive symbols of Ω and an additional primitive symbol $\bar{p}$.

The terms are defined in $\bar{\Omega}$ in the same way as in Ω.

The atomic formulas of the system $\bar{\Omega}$ consist of all the atomic formulas of Ω and all the expressions $\bar{p}\ t$, where t is a term.

The formulas of $\bar{\Omega}$ are defined in the usual way.

The <u>rules</u> of $\bar{\Omega}$ consist of the rules of Ω and the additional rule (which we shall call the S rule):

$$\begin{array}{c} F\ \bar{p}\ t \\ \swarrow \quad \swarrow \quad \searrow \\ F\ \bar{p}\ 0 \quad \ldots \quad F\ \bar{p}(z) \end{array}$$

where t has a numerical value z' and the branches are to <u>all</u> the formulas $F\ \bar{p}\ z$ for $z < z'$

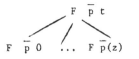

The <u>closure conditions</u> of $\bar{\Omega}$ consist of the closure conditions of Ω together with the two additional conditions:

c) A branch is closed if it contains two formulas $F\,\bar{p}\,s$ and $T\,\bar{p}\,t$, where s and t are numerical terms such that $s \leq t$.

d) A branch is closed if it contains the formula $F\,\bar{p}\,t$ where t is a numerical term with the numerical value 0.

It will turn out that every formula $\bar{p}\,t$ is derivable in $\bar{\Omega}$, but only with a certain least order.

We can prove the two reduction propositions for the cut elimination in the same way as for the system Ω (see §4) except that we must consider two new cases in the first proposition that yields the theorem. The parts to be added to the proof of proposition 21 are the following:

iii. Let one of the branches close by closure condition d. Thus the formula closing the branch is $F\,\bar{p}\,t$ where t has numerical value 0. The other branch must thus close by closure condition c, i.e., it has $T\,\bar{p}\,t$ on its branch as cut formula and also a formula $F\,\bar{p}\,s$, where s is numerically less than or equal to t. But since t is numerically equal to 0, s also has numerical value 0, and the branch closes without need of the cut formulas.

iv. Let both of the branches close by closure condition c. Let $\bar{p}\,s$ be the cut formula. Then one branch contains two formulas $F\,\bar{p}\,s$ and $T\,\bar{p}\,t$ where $s \leq t$ and the other branch contains two formulas $F\,\bar{p}\,r$ and $T\,\bar{p}\,s$ where $r \leq s$. In the resulting branch the formulas $F\,\bar{p}\,s$ and $T\,\bar{p}\,s$ are missing, but since $r \le s$ and $s \le t$ implies $r \le t$, the branches still close without these formulas.

From the two reduction propositions we have as before that any proof with cuts of degree n and rank $\le \alpha$ can be transformed into a primitive proof of rank $\le 2_{n+1}(\alpha)$ of the same formula. If α is smaller than an ϵ-number ϵ, by statement 14 of the introduction, it is also true that $2_{n+1}(\alpha) < \epsilon$. We thus have:

<u>Proposition 26</u>. Every proof with cuts in $\bar{\Omega}$ whose rank is smaller than an ϵ-number ϵ can be transformed into a cut-free derivation in $\bar{\Omega}$ of the same formula with rank less than ϵ.

<u>Proposition 27</u>. Any cut-free derivation of $\bar{p}\,z_1$ in $\bar{\Omega}$ has rank $\ge z_1$.

The proof is by transfinite $<$-induction on z_1.

a) For $z_1 = 0$, the proposition is trivial.

b) Let z_1 not equal 0 and let the proposition hold for all numerals $z < z_1$. Since $z_1 \ne 0$, $F\,\bar{p}\,z_1$ cannot itself cause its branch to close. The only inference rule that can be applied to it is the S-rule. So the tableau for $\bar{p}\,z_1$ must begin:

$$F\,\bar{p}\,z_1$$

$$F\,\bar{p}\,0 \quad \ldots \quad F\,\bar{p}\,z \qquad \text{for all } z < z_1$$

Now by assumption a closed tableau starting with $F\,\bar{p}\,z$ must have rank $\ge z$ for $z < z_1$. Thus the rank of the proof of $\bar{p}\,z_1$ must be $\ge z_1$.

<u>Proposition 28</u>. There is a proof of rank ≤ 6 in $\overline{\Omega}$ for the $<$-progressiveness

$$P_x \overline{p}\, x : \quad (\forall x)((\forall y)(y < x \supset \overline{p}\, y) \supset \overline{p}\, x)$$

of $\overline{p}$.

Proof. The following is a proof of the formula in question.

1 $F\ (\forall x)((\forall y)(y < x \supset \overline{p}\, y) \supset \overline{p}\, x)$

$\delta;1$ (a branch for every numeral z)

$(z = 0)$

2 $F(\forall y)(y < 0 \supset \overline{p}\, y) \supset \overline{p}\, 0$

$\alpha;2$

3 $F\ \overline{p}\ 0$

$\#_d;3$

$(z > 0)$

4 $F(\forall y)(y < z \supset \overline{p}\, y) \supset \overline{p}\, z$

$\alpha;4$

5 $T(\forall y)(y < z \supset \overline{p}\, y)$

$\alpha;4$

6 $F\ \overline{p}\ z$

$S;\ 6$ (a branch for every numeral $z_1 < z$)

$(z_1 = 0)$

7 $F\ \overline{p}\ 0$

$\#_d;7$

$(0 < z_1 < z)$

8 $F\ \overline{p}\ z_1$

$\gamma;5$

9 $T\ z_1 < z \supset \overline{p}\ z_1$

$\beta;9$

10 $F\ z_1 < z$

$\#_a;10$

11 $T\ \overline{p}\ z_1$

$\#_c;8,\ 11$

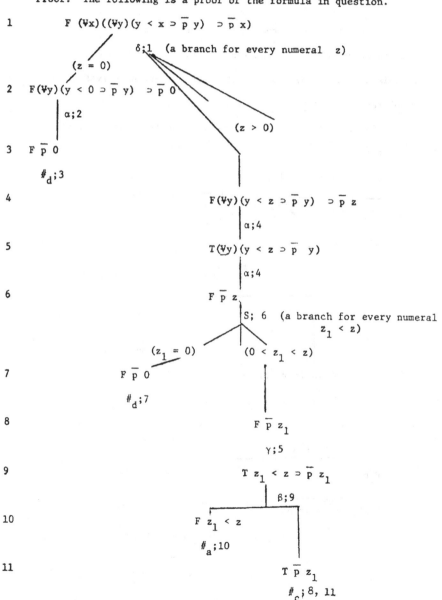

It can be seen by inspection that this proof has rank ≤ 6.

For any one-place predicate variable p we have:

<u>Theorem 12</u>. If ϵ is an ϵ-number and z is a numeral with $\epsilon \leq z$, a cut-free derivation of $J_x(p\,x\,,z)$ in Ω has rank $\geq \epsilon$.

Proof. Let a cut-free derivation of $J_x(p\,x\,,z)$ of rank $\leq \alpha$ be given in Ω. If one replaces all occurrences of p in this derivation by $\bar{p}$, one obtains a cut-free derivation in $\bar{\Omega}$ of rank $\leq \alpha$ for $J_x(\bar{p}\,x\,,z)$.

Using this derivation one can form the following derivation of $\bar{p}\,z$:

1 $F\,\bar{p}\,z$

S;1 (a branch for every numeral $z_1 < z$)

2 $F\,\bar{p}\,0$

$\#_d$;2

3 $F\,\bar{p}\,z_1$

CUT

4 $F\,J_x(\bar{p}\,x\,,z)$

has a closed tree of rank $\leq \alpha$ by assumption

5 $T\,J_x(\bar{p}\,x\,,z)$

AB;5

6 $T\,P_x\bar{p}\,x \supset (\forall x)(x < z \supset \bar{p}\,z)$

β;6

7 $F\,P_x\bar{p}\,x$

has a closed tree of rank ≤ 6 by proposition 28

8 $T(\forall x)(x < z \supset \bar{p}\,z)$

γ;8

(1)

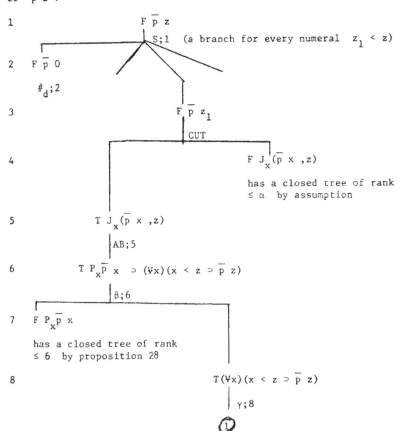

(1)

9 T $z_1 < z \supset \bar{p}\, z$

β;9

10 F $z_1 < z$ — $\#_a$;10

11 T $\bar{p}\, z$ — $\#_b$;1,11

It may be seen from inspection that this proof has rank $\leq \max(\alpha, 8) + 1$. By the cut elimination proposition, proposition 26, since this number would be less than ϵ if α were, the rank of the proof after cuts were removed would also be less than ϵ. Since $\epsilon \leq z$, this would contradict proposition 27. It follows that the assumed derivation of $J_x(p\ x\ , z)$ has rank $\geq \epsilon$.

We now can sharpen theorems 9 and 10:

Theorem 13. The induction formula $J_x(A(x), z)$ is provable for all formulas $A(t)$ in Ψ (Ψ_α) iff z precedes the first ϵ-number ϵ_0 (the first ϵ-number following α).

Proof. I. By theorems 9 and 10 there are derivations of $<$-induction up to the given numbers.

II. a) If $J_x(p\ x\ , z)$ is derivable in Ψ for a one-place predicate variable p, by theorem 8 the formula has a primitive derivation in Ω with a rank $< \epsilon_0$. By theorem 12, $z < \epsilon_0$ must also hold. The induction formula $J_x(A(x), z)$ is thus provable in Ψ for all $A(t)$ only when $z < \epsilon_0$.

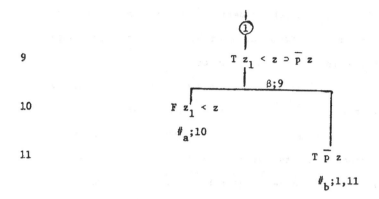

b) If $J_x(p\ x\ ,z)$ is derivable in Ψ_α and ϵ is the first ϵ-number following α, we first extend the appropriate branches of this proof using primitive derivations in Ω of the formulas $J_x(F(x),\alpha)$, which exist with rank $< \epsilon$ by theorem 11. We transform this to a proof (with cuts) in Ω for $J_x(p\ x\ ,z)$ of rank $< \epsilon$. By the cut-elimination theorem, we can eliminate the cuts from this derivation without raising the rank to ϵ. By Theorem 12 it follows that $z < \epsilon$ must hold.

Transfinite induction over other well-ordering relations. The induction formulas considered above are based only on the $<$-induction for the specific ordering relation $<$. For certain other ordering relations $\tilde{<}$, the $\tilde{<}$-induction can be shown to follow from the $<$-induction. We make the following assumptions about the ordering relation $\tilde{<}$:

1. The relation $\tilde{<}$ is a decidable relation for the numerals of the system Ψ.

There exist evaluable functions of z, $\tilde{z}$ and $\rho(z)$, such that:

2. For every z_1 and z_2, $z_1 < z_2$ is equivalent to $\tilde{z}_1 \tilde{<} \tilde{z}_2$.

3. For every numeral z_1 for which there is a numeral z_2 with $z_1 \tilde{<} \tilde{z}_2$, there is a numeral $\rho(z_1)$ with $\widetilde{\rho(z_1)} = z_1$. (Here the ordinal numbers represented by the $<$-relation are mapped onto a cut of the ordinal numbers represented by the $\tilde{<}$-relation.)

Thus in Ψ the following are arithmetically true formulas:

(1) $$b \tilde{<} \tilde{a} \supset \widetilde{\rho(b)} = b,$$

(2) $$b \tilde{<} \tilde{a} \supset \widetilde{\rho(b)} < a$$

We define $\tilde{<}$-progressiveness and $\tilde{<}$-induction as we did $<$-progressiveness and $<$-induction except that we change the symbol for the well-ordering.

<u>Lemma</u>. The formula $J_x(A(\tilde{x}),a) \supset \tilde{J}_x(A(x),\tilde{a})$ is provable in Ψ.

Proof. The following tableau comprises a proof.

1 $$F\ J_x(A(\tilde{x}),a) \supset \tilde{J}_x(A(x),\tilde{a})$$

α;1

2 $$T\ J_x(A(\tilde{x},a)$$

α;1

3 $$F\ \tilde{J}_x(A(x),\tilde{a})$$

AB;2

4 $$T\ P_xA(\tilde{x}) \supset (\forall y)(y < a \supset A(\tilde{y}))$$

AB;3

5 $$F\ \tilde{P}_xA(x) \supset (\forall y)(y \tilde{<} \tilde{a} \supset A(y))$$

α;5

6 $$T\ \tilde{P}_xA(x)$$

α;5

7 $$F(\forall y)(y \tilde{<} \tilde{a} \supset A(y))$$

AB;6

8 $$T(\forall x)((\forall y)(y \tilde{<} x \supset A(y)) \supset A(x))$$

δ;7

(1)

(1)

9 F $b \tilde{<} \tilde{a} \supset A(b)$

α;9

10 T $b \tilde{<} \tilde{a}$

α;9

11 F $A(b)$

AX;(1)

12 T $b \tilde{<} \tilde{a} \supset \widetilde{\rho(b)} = b$

AX;(2)

13 T $b \tilde{<} \tilde{a} \supset \rho(b) < a$

β;12

14 F $b \tilde{<} \tilde{a}$

$\#_b$;10, 14

15 T $\widetilde{\rho(b)} = b$

β;13

16 F $b \tilde{<} \tilde{a}$

$\#_b$;10, 16

17 T $\rho(b) < a$

β;4

18 T$(\forall y)(y < a \supset A(\tilde{y}))$

γ;18

19 T $\rho(b) < a \supset A(\widetilde{\rho(b)})$

β;19

20 F $\rho(b) < a$

$\#_b$;17,20

21 T $A(\widetilde{\rho(b)})$

$\$_c$;11,15,21

(2)

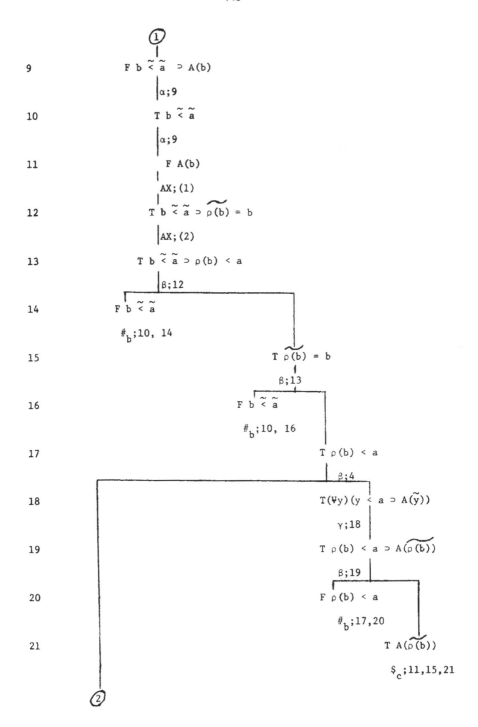

(2)

22 $F \; P_x A(\tilde{x})$

AB; 22

23 $F(\forall x)((\forall y)(y < x \supset A(\tilde{y})) \supset A(\tilde{x}))$

δ; 23

24 $F(\forall y)(y < c \supset A(\tilde{y})) \supset A(\tilde{c})$

α; 24

25 $T(\forall y)(y < c \supset A(\tilde{y}))$

α; 24

26 $F \; A(\tilde{c})$

γ; 8

27 $T(\forall y)(y \,\tilde{<}\, \tilde{c} \supset A(y)) \supset A(\tilde{c})$

β; 27

28 (right branch:) $T \; A(\tilde{c})$

$\$_b$; 26, 29

29 $F(\forall y)(y \,\tilde{<}\, \tilde{c} \supset A(y))$

δ; 29

30 $F \; d \,\tilde{<}\, \tilde{c} \supset A(d)$

α; 30

31 $T \; d \,\tilde{<}\, \tilde{c}$

α; 30

32 $F \; A(d)$

γ; 25

33 $T \; (\rho(d) < c \supset A(\widetilde{\rho(d)}))$

AX; (1)

(3)

(3)

34 T $d \tilde{<} \tilde{c} \supset \widetilde{\rho(d)} = d$

AX; (2)

35 T $d \tilde{<} \tilde{c} \supset \rho(d) < c$

β; 34

36 F $d \tilde{<} \tilde{c}$

$\#_b$; 31,36

37 T $\widetilde{\rho(b)} = d$

β; 35

38 F $d \tilde{<} \tilde{c}$

$\#_b$; 31,38

39 T $\rho(d) < c$

β; 33

40 F $\rho(d) < c$

$\#_b$; 39,40

41 T $A(\widetilde{\rho(d)})$

$\$_c$; 32,37,41

From the lemma we obtain:

Proposition 29. If <-induction is derivable in Ψ (Ψ_α) for every formula F(t) over numerals < z, $\tilde{<}$-induction is also derivable for an arbitrary formula A(t) over numerals $\tilde{<} \tilde{z}$.

The ω-incompleteness. Let p be a one-place predicate variable and define

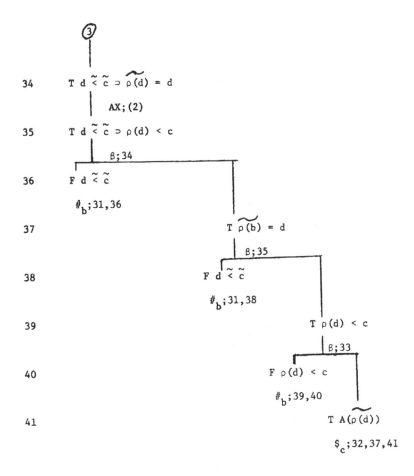

$$U(t) \equiv P_x p\ x \supset (t < \epsilon_0 \supset p\ t).$$

Then we have:

<u>Theorem 14</u>. (ω-incompleteness of the system Ψ). In Ψ, $U(z)$ is provable for every numeral z, but $(\forall y)U(y)$ is not provable.

Proof. I. a) Let z be numeral $< \epsilon_0$. Then $z^* < \epsilon_0$ also and by theorem 9 $J_x(p(x), z^*)$ is derivable in Ψ. With this theorem we can form the following proof for $U(z)$:

1 $F\ U(z)$

AB;1

2 $F\ P_x p\ x \supset (z < \epsilon_0 \supset p\ z)$

α;2

3 $T\ P_x p(x)$

α;2

4 $F\ z < \epsilon_0 \supset p(z)$

α;4

5 $F\ p\ z$

TH (this is $J_x(p\ x, z^*)$)

6 $T\ P_x p\ x \supset (\forall x)(x < z^* \supset p\ x)$

β;6

7 $F\ P_x p\ x$ (left branch)

$\$_b$;3,7

8 $T(\forall x)(x < z^* > p\ z)$ (right branch)

γ;8

①

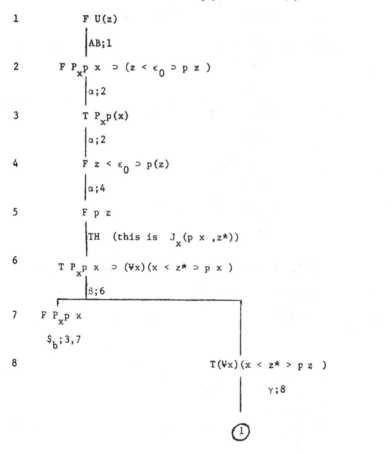

(1)

9 $T\ z < z^* \supset p\ z$

β;9

10 $F\ z < z^*$ | 11 $T\ p\ z$

$\#_a$;10 | $\#_b$;5,11

b) Let z be a numeral such that $z \geq \epsilon_0$. We then have the following proof of $U(z)$:

1 $F\ U(z)$

AB;1

2 $F\ P_x p\ x \supset (z < \epsilon_0 \supset p\ z)$

α;2

3 $F\ z < \epsilon_0 \supset p\ z$

α;3

4 $T\ z < \epsilon_0$

$\#_a$;4

II. If $(\forall y)U(y)$ were derivable in Ψ, we would have the following derivation of $J_x(p\ x, \epsilon_0)$ in Ψ:

1 $F\ J_x(p\ x, \epsilon_0)$

AB;1

2 $F\ P_x p\ x \supset (\forall x)(x < \epsilon_0 \supset p\ x)$

TH (by assumption)

3 $T(\forall y)U(y)$

AB;3

(1)

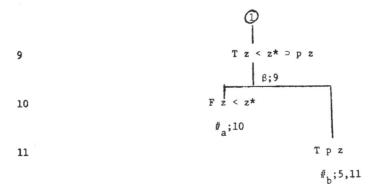

(1)

4 $T(\forall y)(P_x p\ x \supset (y < \epsilon_0 \supset p\ y\ )$

$\alpha;2$

5 $T\ P_x p\ x$

$\alpha;2$

6 $F(\forall x)(x < \epsilon_0 \supset p\ x\ )$

$\delta;6$

7 $F\ a < \epsilon_0 \supset p\ a$

$\gamma;4$

8 $T\ P_x p\ x \supset (a < \epsilon_0 \supset p\ a\ )$

$\beta;8$

9 $F\ P_x p\ x$

$\$_b;5,9$

10 $T\ a < \epsilon_0 \supset p\ a$

$\$_b;7,10$

Since we know by theorem 13 that no such proof can exist in Ψ, $(\forall y)U(y)$ is not provable in Ψ.

In order to show the ω-incompleteness of the system Ψ_α we use the formula

$$U_\alpha(t) \equiv P_x p\ x \supset (t < \epsilon \supset p\ t\ )$$

where ϵ is the first ϵ-number following α. One proves just as above:

Theorem 15. (ω-incompleteness of the system Ψ_α). In Ψ_α the formula $U_\alpha(z)$ is provable for every numeral z, but $(\forall y)U_\alpha(y)$ is not provable.

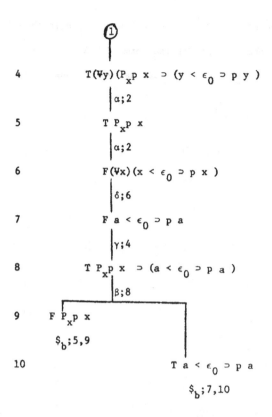

If one extends the system Ψ to a system even stronger than the Ψ_α by adding all the formulas of the form $F\ P_x A(x) \supset A(t)$ in the closure conditions, the ω-incompleteness can no longer be shown with the help of $<$induction.

CHAPTER II - SECOND ORDER LOGIC

Recently, many proof-theoretists have been working on the problem of finding a constructive consistency proof for second order logic and type theory. Since the Hauptsatz (cut elimination theorem) provided this for first order logic and first order number theory, this was one of the first approaches considered. It was not immediately evident that the theorem was true in the higher order cases, but Tait [1] and Prawitz [1] finally found proofs for second order logic and then Prawitz [2] and Takahashi [1] found proofs for type theory. These proofs were non-constructive, however, and thus did not satisfy the proof-theoretists' goal. They were of interest, nevertheless, not only heuristically in indicating that this still might be a fruitful direction to consider, but also in throwing some light on the significance of the cut rule in the higher-order logics, where even cut-free proofs do not have a simple subformula property as do cut-free proofs in first order logic or first order number theory. This chapter is devoted to further exploring that subject and others related to it.

We will consider a sequence of theorems with syntactic content, a constructive proof of any one of which would yield a constructive consistency proof. It is interesting to see how the classical semantic proofs that we give differ greatly in complexity, the "Hauptsatz" theorem having the most difficult proof. We will base most of our discussion on formal tableau systems in which comprehension axioms are used. These systems were chosen for their simplicity, with the hope that new insight might be obtained into syntactic questions.

It seems that the hope has been justified to a certain extent, as will be made more specific in §3. (To the author's knowledge, no system very closely related to this has been used in higher order logic.)

We will carry out our whole discussion in terms of second order logic. It seems that all the results extend to higher order logics and type theory, but unpleasant complications arise at almost every step. For simplicity also we will usually restrict ourselves to speaking about proofs of single formulas rather than of tableaux proving the inconsistency of sets of formulas.

§1. Two Formulations of Second Order Logic; Translation Procedures

Most of the recent work on second order logic and type theory has been in terms of Gentzen- or Schütte-type systems with abstraction terms. It is with respect to such a system that a Hauptsatz has been formulated and proved non-constructively. I will now describe the associated tableau systems in terms of which all the recent results and proofs can be translated in the most direct manner. I will call the cut-free system A_1^2, where the 2 indicates a second order system, and the A indicates that abstraction terms are used.

The System A_1^2. The primitive symbols are:

1) a denumerable list of symbols $a_1, \ldots,$ called individual parameters;

2) a denumerable list of symbols $x_1, \ldots,$ called individual variables;

3) for each integer $n > 0$, a denumerable list of symbols P_1, ..., called _n-ary predicate parameters_;

4) for each integer $n > 0$, a denumerable list of symbols Z_1, ..., called _n-ary predicate variables_;

5) the logical symbols $\neg$, $\wedge$, $\vee$, $\supset$, $\forall$, $\exists$;

6) the abstraction operator λ;

7) a comma and parentheses.

We will use c^0 (z^0) to denote an individual parameter (variable), and c^n (z^n) with $n > 0$ to denote an n-ary predicate parameter (variable). We sometimes call an individual parameter or variable a 0-ary parameter or variable. We will use y^n with $n \geq 0$ to denote a symbol that may be either a variable z^n or a parameter c^n. We will attach subscripts to symbols when necessary.

We define the _abstraction terms_ and _formulas_ of the system A_1^2 simultaneously as follows:

1) a string of the form $y^n c_1^0, \ldots, c_n^0$ (where $n > 0$) is an _atomic formula_ and a formula;

2) if A is a formula, so is $\neg A$;

3) if A and B are formulas, so are $(A \wedge B)$, $(A \vee B)$, $(A \supset B)$;

4) if A is a formula, $(\exists z^n)A$ and $(\forall z^n)A$ are formulas (where $n \geq 0$);

5) if A is a formula and $n > 0$, $\lambda x_1 \ldots x_n A$ is an n-ary abstraction term;

6) if t^n is an n-ary abstraction term, $t^n c_1^0, \ldots, c_n^0$ is a formula.

An n-ary term (a term) is a parameter c^n $(n>0)$ or an n-ary abstraction term. We will use t^n to denote such a term. We abbreviate $((A \supset B) \wedge (B \supset A))$ by $(A \leftrightarrow B)$ and $(\forall z_1)\dots(\forall z_n)$ by $(\forall z_1 \dots z_n)$. Substitution, closed formulas and signed formulas are defined as in first order logic (of course, here a closed formula must also have each of its predicate variables within the scope of some quantifier), and, as usual, we will use the word "formula" to mean "closed signed formula" unless mention is made to the contrary or the context makes it clear that another meaning is intended. We will use γ^n to denote the formulas $T(\forall z^n)A(z^n)$ and $F(\exists z^n)A(z^n)$ and δ^n to denote $F(\forall z^n)A(z^n)$ and $T(\exists z^n)A(z^n)$. When the kind of the first variable in a γ^n or δ^n formula is irrelevant or understood, we will sometimes speak of a γ or δ formula, (or a γ or δ rule - see below). We call a formula $\pi\ \lambda x_1 \dots x_n\ Ac_1^0, \dots, c_n^0$ (or its body) a λ formula. Now in our tableau system A_1^2 the rules are:

α rule
$$\begin{array}{cc} \alpha & \alpha \\ | & | \\ \alpha_1 & \alpha_2 \end{array}$$

β rule
$$\begin{array}{ccc} & \beta & \\ / & & \backslash \\ \beta_1 & & \beta_2 \end{array}$$

γ^n rule
$$\begin{array}{c} \gamma^n \\ | \\ \gamma^n(t^n) \end{array}$$

δ^n rule
$$\begin{array}{c} \delta^n \\ | \\ \delta^n(c^n) \end{array} \quad , c^n \text{ new}$$

λ rule
$$\begin{array}{c} \pi\ \lambda x_1 \dots x_n\ A(x_1, \dots, x_n)c_1, \dots, c_n \\ | \\ \pi A(c_1, \dots, c_n) \end{array}$$

A branch of a tableau is closed when it contains an atomic formula and its conjugate.[1]

---

[1] We can prove by induction on the number of symbols in a formula that a branch containing any formula and its conjugate either is or can be extended to a closed branch.

We will use A_5^2 to denote the system that is like A_1^2 except that the <u>cut rule</u>

$$\begin{array}{ccc} & X & \\ \swarrow & & \searrow \\ Y & & \overline{Y} \end{array}$$

is added.[2] (We will often speak of A^2 formulas, because the formulas of A_1^2 and A_5^2 are the same.) The Hauptsatz proved by Tait, Prawitz and Takahashi says that any formula that can be proved with the use of the cut rule can be proved without its use, i.e., that the systems A_1^2 and A_5^2 are equivalent. A result of Takeuti [1] shows that this implies the consistency of second order arithmetic.

We will now consider a system equivalent to A_5^2, which we call C_5^2, the 2 again indicating a second order system, the C indicating that comprehension axioms are being used.

<u>The System</u> C_5^2. The symbols of C_5^2 are the same as those of A_5^2, except that the abstraction operator is not used. <u>Formulas</u> are defined in C_5^2 in the same way as in A_5^2 except that abstraction terms are not used. Notational conventions will be the same when applicable.

Let $A(z_1,\dots z_i, x_1,\dots x_n)$ be a formula of C_5^2 without parameters, all of whose variables are indicated. Then

$$(\forall z_1\dots z_i)(\exists Z)(\forall x_1\dots x_n)(Zx_1\dots x_n \leftrightarrow A(z_1,\dots,z_i, x_1,\dots,x_n))$$

---

[2] The significance of the lower subscript will become clear later in §3.

is an axiom, or, more precisely, an n-ary comprehension axiom. Here $0 \le i$, $0 < n$, but the variables denoted are not actually required to occur in $A(z_1,\ldots,z_i, x_1,\ldots,x_n)$. We use ζ^n or ζ to denote an n-ary comprehension axiom preceded by the prefix T, which we also call an (n-ary) axiom.

Now the tableau rules used in C_5^2 are as follows:

α rule

$$\begin{array}{c} \alpha \\ | \\ \alpha_1 \end{array} \qquad \begin{array}{c} \alpha \\ | \\ \alpha_2 \end{array}$$

β rule

$$\begin{array}{ccc} & \beta & \\ \swarrow & & \searrow \\ \beta_1 & & \beta_2 \end{array}$$

γ^n rule

$$\begin{array}{c} \gamma^n \\ | \\ \gamma^n(c^n) \end{array}$$

δ^n rule

$$\begin{array}{c} \delta^n \\ | \\ \delta^n(c^n) \end{array} \quad , \; c^n \text{ new}$$

axiom rule

$$\begin{array}{c} X \\ | \\ \zeta \end{array}$$

cut rule

$$\begin{array}{ccc} & X & \\ \swarrow & & \searrow \\ Y & & \overline{Y} \end{array}$$

Thus we see that the rules of C_5^2 differ from those of A_5^2 in that (1) in the γ^n rule only parameters can be substituted, (2) the axiom rule is added, (3) there is no λ rule. Again a branch of a tableau is closed when it contains an atomic formula and its conjugate.

We note that in the Gentzen system corresponding to this tableau system, the rules are those of a two-sorted predicate calculus except that comprehension axioms are allowed to be eliminated from the left sides of sequents. Also, let us define the A^2 formula $X^{\mathcal{T}}$ associated with any C_5^2 formula X with respect to a C_5^2 tableau $\mathcal{T}$ as follows: We first let the level of a parameter P^m introduced through an axiom in $\mathcal{T}$ be the level at which

$$T(\forall x_1 \ldots x_m)(P^m x_1,\ldots,x_m \leftrightarrow A(a_1,\ldots,a_2, x_1,\ldots,x_m))$$

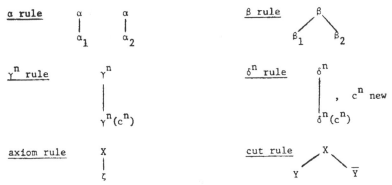

occurs. Let us first substitute for each such parameter P^m of highest level in X the expression $\lambda x_1 \ldots x_m\ A(a_1,\ldots,a_i,\ x_1,\ldots,x_m)$. Clearly the resulting A^2 formula only has parameters of lower level. We continue to make such substitutions until the resulting A^2 formula has no parameters introduced by an axiom in $\mathcal{T}$, and call this formula $X^{\mathcal{T}}$. We can show by induction on the number of symbols in $X^{\mathcal{T}}$ that:

Proposition 30. Any closed branch of a C_5^2 tableau $\mathcal{T}$ containing a formula X and $\overline{X}$ can be extended to a branch satisfying the following condition: the branch contains an atomic formula and its conjugate where the predicate parameter of the formula was not introduced through an axiom (i.e. by the first δ rule application in an axiom's descendents on a branch).

We now note that the cut rule

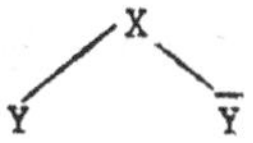

is a derivable rule of the system C_5^2:

Proposition 31. Any proof of C_5^2 in which the cut rule appears can be transformed into a cut-free proof of the same formula in C_5^2.

Proof. We can transform a proof in C_5^2 with cuts into one without cuts by changing only the parts of the proof where the cuts occur.

We will first describe an abbreviation for a part of a tableau we will use in systems with comprehension axioms. Let $A(z_1,\ldots,z_i,\ x_1,\ldots,x_n)$ be an open formula without parameters, all of whose variables are indicated. Let P^n be an n-ary predicate parameter $(n > 0)$, $c_1,\ldots,c_i$ be parameters of the same kind as the respective variables $z_1,\ldots,z_i$. Then when

$$\Big|\,\text{AX}$$

$$P^n\text{: } A(c_1,\ldots,c_i,\ x_1,\ldots,x_n)$$

occurs in a tableau, it will be an abbreviation for

$$\Big|\,\zeta$$

$$T(\forall z_1\ldots z_i)(\exists Z)(\forall x_1\ldots x_n)(Zx_1\ldots x_n \leftrightarrow A(z_1,\ldots,z_i,\ x_1,\ldots,x_n))$$

$$\Big|\,\gamma$$

$$T(\forall z_2\ldots z_i)(\exists Z)(\forall x_1\ldots x_n)(Zx_1\ldots x_n \leftrightarrow A(c_1,z_2,\ldots,z_i,\ x_1,\ldots,x_n))$$

$$\vdots\ \gamma$$

$$T(\exists Z)(\forall x_1\ldots x_n)(Zx_1,\ldots,x_n \leftrightarrow A(c_1,\ldots,c_i,\ x_1,\ldots,x_n))$$

$$\Big|\,\delta^n$$

$$T(\forall x_1\ldots x_n)(P^n x_1,\ldots,x_n \leftrightarrow A(c_1,\ldots,c_i,\ x_1,\ldots,x_n))$$

We will speak of this as an <u>axiom sequence</u> for $A(c_1,\ldots,c_i,\ x_1,\ldots,x_n)$ and P^n.

Now let us describe how we can transform a tableau $\mathcal{T}$ with cuts to one without them. So let a cut

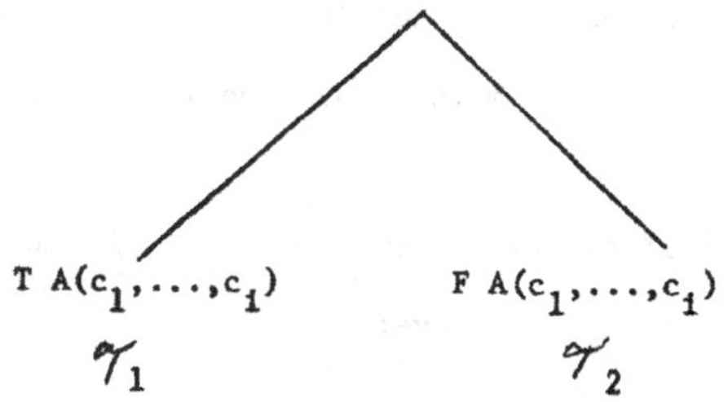

occur in $\mathcal{T}$, where all the parameters of the cut formula are indicated. Let P be an unary predicate parameter and c be an individual parameter new to the tableau. We transform the given part of the tableau to

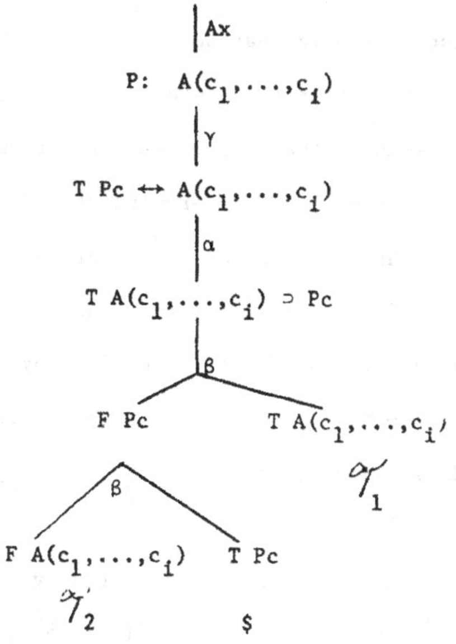

Thus we see that in removing a cut from a proof we replace the cut with the use of an axiom whose purpose is merely to introduce the cut formula.[3]

If we now call the system that is like C^2_5 except for not having the cut rule C^2_4, we have:

<u>Corollary</u>. The systems C^2_4 and C^2_5 are (constructively) equivalent.

(Again, in the future we will speak of C^2 formulas, because all the systems C^2_i will share the same formulas.)

Now we first wish to see in what sense A^2_5 and C^2_5 are equivalent. We notice immediately that the systems differ not only in logical structure, but also in their formulas - the formulas of A^2_5 that contain abstraction terms are not formulas of C^2_5, although all the formulas of C^2_5 are formulas of A^2_5. For any formula of A^2_5, there are formulas of C^2_5 that seem to "mean" the same thing, however. It would be natural to try to define for any formula E of A^2_5 a corresponding formula E' of C^2_5 with the same meaning and then show that E is provable in A^2_5 iff E' is provable in C^2_5.

Let us define such a correspondence inductively:

1) if E is $y^n c^0_1, \ldots, c^0_n$, then E' is also $y^n c^0_1, \ldots, c^0_n$;

[3]It may appear that proving this proposition requires the use of an axiom where the individual variables after the predicate on the left-hand side of the arrow do not occur on the right-hand side of the arrow. A slightly more complicated proof can be given easily, however, even in the case when such axioms are not used.

2) if E is $\neg A$, then E' is $\neg (A)'$;

3) if E is $(A \wedge B)$, $(A \vee B)$, or $(A \supset B)$, then E' is respectively $(A' \wedge B')$, $(A' \vee B')$, $(A' \supset B')$;

4) if E is $(\exists z^n)A$ or $(\forall z^n)A$, then E' is $(\exists z^n)A'$ or $(\forall z^n)A'$ respectively;

5) if E is $\lambda x_1 \ldots x_n A\ (x_1, \ldots, x_n) c_1^0, \ldots, c_n^0$, then E' is $A'(c_1^0, \ldots, c_n^0)$.

We will sometimes call E' the <u>translate</u> of E. Also if Y is a signed formula πE the <u>translate</u> Y' of Y will be $\pi E'$.

We notice that the following are true for any A^2 formula X:

a. If X is an α and $(X')_i = (X_i)'$.

b. If X is a β, X' is a β and $(X')_i = (X_i)'$.

c. If X is a γ, X' is a γ and $(\gamma')(c) = (\gamma(c))'$.

d. If X is a δ, X' is a δ and $(\delta')(c) = (\delta(c))'$.

Now we can show rather easily:

<u>Theorem 16</u>. If E is provable in A_5^2, E' is provable in C_5^2.

Proof. We will give a <u>constructive</u> proof of this theorem, actually showing how to obtain from a proof $\mathcal{P}$ of E in A_5^2 a proof of E' in C_5^2. We note that any cut in our resulting proof will be obtained from a cut in the original proof.

We transform $\mathcal{P}$ in five steps as follows:

1) For each use of the γ^n rule to substitute an abstraction term t^n for a variable, we choose a distinct new parameter P^n and substitute P^n throughout the tableau for all those occurrences of the term t^n that resulted from its introduction by the given use of the rule. We call the result $\mathcal{P}^*$ and note that all A_5^2 rule applications are still valid except for certain λ rule uses where such a

term t^n occurred in the premise formula, and now the premise is an atomic formula with a predicate parameter introduced by step 1.

2) We take the translate of every formula in $\mathcal{P}^*$ to form $\mathcal{P}^{*\prime}$. This time all the legal A_5^2 rule applications of $\mathcal{P}^*$ go into legal C_5^2 rule applications on $\mathcal{P}^{*\prime}$ or to repetitions.

3) Just above each formula $\gamma^n(P^n)$ in $\mathcal{P}^{*\prime}$ containing the highest occurrence of a parameter P^n introduced in step 1) as a replacement for an abstraction term

$$\lambda x_1 \ldots x_n A(a_1,\ldots,a_i,\ x_1,\ldots,x_n)$$

we insert the axiom sequence

$$P^n\colon\ A(a_1,\ldots,a_i,\ x_1,\ldots,x_n)'.$$

We call the result $\mathcal{Q}^{*\prime}$.

4) Finally, let Y be a formula

$$\pi\ A(a_1,\ldots,a_i,\ c_1,\ldots,c_n)$$

derived from

$$\pi\ \lambda x_1 \ldots x_n\ A(a_1,\ldots,a_i,\ x_1,\ldots,x_n)c_1,\ldots,c_n$$

by a λ rule in $\mathcal{P}$, where P^n replaces $\lambda x_1 \ldots x_n\ A(a_1,\ldots,a_i,\ x_1,\ldots,x_n)$ in $\mathcal{P}^*$. We change $\mathcal{Q}^{*\prime}$ as follows so that the image of Y,

$$\pi\ A(a_1,\ldots,a_i,\ c_1,\ldots,c_n)',$$

in it is obtained by a legal C_5^2 rule:

a) if π is T we replace the image of Y in $\mathcal{Q}^{*\prime}$ by

$$\Big|\,\gamma$$

$$T(\forall x_2 \ldots x_n)(P^n c_1, x_2, \ldots, x_n \leftrightarrow A(a_1, \ldots, a_i, c_1, x_2, \ldots, x_n)')$$

$$\vdots\ \gamma$$

$$T\ P^n c_1, \ldots, c_n \leftrightarrow A(a_1, \ldots, a_i, c_1, \ldots, c_n)'$$

$$\Big|\,\alpha$$

$$T\ P^n c_1, \ldots, c_n \supset A(a_1, \ldots, a_i, c_1, \ldots, c_n)'$$

$$\beta$$

$$F\ P^n c_1, \ldots, c_n \qquad\qquad T\ A(a_1, \ldots, a_i, c_1, \ldots, c_n)'$$

$$\$$$

which we call the <u>T-expansion tree</u> for the axiom sequence

P^n: $A(a_1, \ldots, a_i, x_1, \ldots, x_n)'$ and $\langle c_1, \ldots, c_n \rangle$;

b) if π is F we replace the image of Y in *' by

$$\Big|\,\gamma$$

$$T(\forall x_2 \ldots x_n)(P^n c_1, x_2, \ldots, x_n \leftrightarrow A(a_1, \ldots, a_i, c_1, x_2, \ldots, x_n)')$$

$$\vdots\ \gamma$$

$$T\ P^n c_1, \ldots, c_n \leftrightarrow A(a_1, \ldots, a_i, c_1, \ldots, c_n)'$$

$$\Big|\,\alpha$$

$$T\ A(a_1, \ldots, a_1, c_1, \ldots, c_n)' \supset P^n(c_1, \ldots, c_n)$$

$$\beta$$

$$F\ A(a_1, \ldots, a_i, c_1, \ldots, c_n)' \qquad T\ P^n(c_1, \ldots, c_n)$$

$$\$$$

which we call the *F-expansion tree* for the axiom sequence P^n: $A(a_1,\dots,a_i,x_1,\dots,x_n)$ and $\langle c_1,\dots,c_n\rangle$. (Here the \$'s mark closed branches.) We call T- or F-expansion trees, *expansion trees*.) We call the bottom element $\pi\, A(a_1,\dots,a_i,c_1,\dots,c_n)$, the *final formula* of a T- or F-expansion tree and call the bottom element $\bar{\pi}\, P^n c_1,\dots,c_n$ the *closing* formula of the expansion subtree.) We call the resulting tableau $\mathcal{T}^*$. We see easily that all formulas in $\mathcal{T}^*$ are obtained by legal C_5^2 rules except for the formulas that are repetitions introduced in step 2).

5) We eliminate repetitions, calling the result $\mathcal{T}$. It remains to see that $\mathcal{T}$ is closed. But any branch in $\mathcal{T}$ is either created as a closed branch in step 4) or comes from a branch in $\mathcal{P}$ that contained an atomic formula and its conjugate. None of the steps 1) - 5) affects atomic formulas, so in the latter case also the branch in $\mathcal{T}$ contains a closure set.

This completes the proof of Theorem 16.

Now we recall that the system A_1^2 was the system A_5^2 without the cut rule. If we examine the above proof of Theorem 16, we see that a cut-free A_5^2 proof (i.e. an A_1^2 proof) goes into a C_5^2 proof satisfying the following properties:

P_1. The cut rule is not used.

P_2. The axiom rule is only used with an axiom sequence

$$P^n\colon A(a_1,\dots,a_i,x_1,\dots,x_n)$$

and only when the next rule applied is a γ^n rule in which P^n is substituted for a variable.

P_3. The last formula

$$T(\forall x_1 \ldots x_n)(P^n x_1,\ldots,x_n \leftrightarrow A(a_1,\ldots,a_i,x_1,\ldots,x_n)$$

of an axiom sequence is

a) used in a <u>T-expansion tree</u>

$$T(\forall x_2 \ldots x_n)(P^n c_1,x_2,\ldots,x_n \leftrightarrow A(a_1,\ldots,a_i,c_1,x_2,\ldots,x_n))$$

$$\vdots$$

$$T\ P^n c_1,\ldots,c_n \leftrightarrow A(a_1,\ldots,a_i,c_1,\ldots,c_n)$$

$$T\ P^n c_1,\ldots,c_n \supset A(a_1,\ldots,a_i,c_1,\ldots,c_n)$$

$$F\ P^n c_1,\ldots,c_n \qquad T\ A(a_1,\ldots,a_i,c_1,\ldots,c_n)$$

only when the formula $T\ P^n c_1,\ldots,c_n$ is above it in the tree;

b) used in a <u>F-expansion tree</u>

$$T(\forall x_2 \ldots x_n)(P^n c_1,x_2,\ldots,x_n \leftrightarrow A(a_1,\ldots,a_i,c_1,x_2,\ldots,x_n))$$

$$\vdots$$

$$T\ P^n c_1,\ldots,c_n \leftrightarrow A(a_1,\ldots,a_i,c_1,\ldots,c_n)$$

$$T\ A(a_1,\ldots,a_i,c_1,\ldots,c_n \supset P^n c_1,\ldots,c_n$$

$$F\ A(a_1,\ldots,a_i,c,\ldots,c_n) \qquad T\ P^n c_1,\ldots,c_n$$

only when the formula $F\ P^n c_1,\ldots,c_n$ is above it in the tree,[4]

and

---

[4] In fact, in the C_5^2 proof we have obtained, the expansion trees always <u>directly</u> follow the conjugate of the closing formula of the tree.

c) is used only in T- or F-expansion trees.[5]

Now if we denote by C_1^2 the system that is like C_5^2 except that the restrictions P_1, P_2, P_3 are imposed, the proof of Theorem 16 also yields a constructive proof of:

Theorem 17. If E is provable in A_1^2, E' is provable in C_1^2.

We can show more, however, namely that the system A_1^2 is actually equivalent to the system C_1^2, and that the system A_5^2 is equivalent to the system C_5^2.

Before we do this we will review some old concepts and introduce some new ones.

We recall that we have six different kinds of signed A^2 formulas: atomic, $\alpha,\beta,\gamma,\delta$, and λ formulas. If X is a λ formula, we let X^λ denote the successor of X under the λ rule, and then if X^λ is a λ formula, we let X^{λ^2} denote the successor of X^λ by the λ rule, etc. If X is a λ formula and i is the largest integer for which X^{λ^i} is a λ formula, we let $X^{\underline{\lambda}}$ denote $X^{\lambda^{i+1}}$ and if X is not a λ formula, we let $X^{\underline{\lambda}}$ be X; also, if $j > i$, we let X^{λ^j} mean $X^{\underline{\lambda}}$, i.e., X^{λ^i}. We can see that:

[5] We note also that all the branches in our C_5^2 proof are closed due to parameters not introduced by axioms.

If X is a λ, $(X^{\lambda^i})' = X'$; in particular, $(X^{\underline{\lambda}})' = X'$.

The following theorem will enable us to shorten later proofs.

<u>Theorem 18</u>. Let A and B be any two A^2 formulas for which A' is the same as B'. Then we can transform any proof $\mathcal{P}$ of A in A_1^2 (A_5^2) into a proof $\mathcal{T}$ of B in A_1^2 $(A_5^2$, respectively).

Proof. Let us be given an A, B and $\mathcal{P}$ satisfying the hypothesis of the theorem.

Let us first transform $\mathcal{P}$ to a tableau $\mathcal{P}^{\dagger}$ as follows: we eliminate all λ rule conclusions in $\mathcal{P}$ and then place after any λ formula z remaining

$$z^{\lambda}$$
$$z^{\lambda^2}$$
$$\vdots$$
$$z^{\underline{\lambda}}$$

i.e., essentially we move the conclusion of a λ rule application up to a position immediately after the premise. This is clearly still a A_1^2 (A_5^2) proof of A.

We can divide $\mathcal{P}^{\dagger}$ into disjoint so-called <u>λ-subpaths</u>:

$$z$$
$$z^{\lambda}$$
$$\vdots$$
$$z^{\underline{\lambda}}$$

where z is not the conclusion of a λ rule. If W is not a λ-formula in $\mathcal{P}^{\dagger}$, we will denote the λ-subpath ending with W by S_W.

If S_W is a λ-subpath, we say the <u>level</u> of S_W is the level of W, the lowest formula in S_W. Let us number the λ-subpaths of $\mathcal{P}^{\dagger}$ so that a) if S_W has a higher level than S_Z, then S_W has a higher number than S_Z, and b) if S_W and S_Z have the same level but S_W occurs to the left of S_Z in the tableau, then S_W has a lower number than S_Z (we will call this the top-to-bottom left-to-right numbering). For instance the λ-subpaths in a tableau might be numbered as follows:

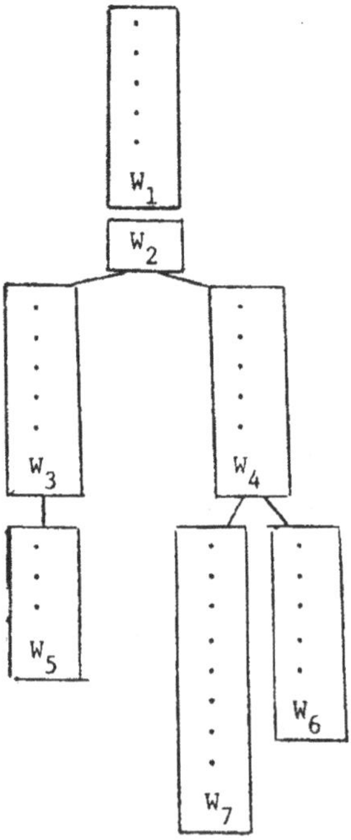

We will construct $\mathcal{T}$ step by step, one step for each λ-subpath of $\mathcal{P}^{\dagger}$. Let $\mathcal{P}^{\dagger n}$ be the part of $\mathcal{P}^{\dagger}$ consisting of the first n λ-subpaths and let us call $\mathcal{T}^n$ the part of $\mathcal{T}$ constructed at step n. We assume the following about $\mathcal{T}^n$ as an induction hypothesis:

$\mathcal{T}^n$ will consist of n disjoint λ-subpaths, numbered top-to-bottom left-to-right, such that if the i-th λ-subpath of $\mathcal{P}^{\dagger n}$ is above the j-th λ-subpath of $\mathcal{P}^{\dagger n}$, the i-th λ-subpath of $\mathcal{T}^n$ is above the j-th λ-subpath of $\mathcal{T}^n$. Also for any $i \leq n$, if W is any member of the i-th λ-subpath in $\mathcal{P}^{\dagger}$ and W* is any member of the i-th λ-subpath in $\mathcal{T}^n$, W' is the same as W*'.

So let us start the construction of $\mathcal{T}^n$. At step 1, we make $\mathcal{T}^1$ the single λ-subpath

$$\begin{array}{c} Y \\ Y^{\lambda} \\ \cdot \\ \cdot \\ \cdot \\ Y^{\underline{\lambda}}. \end{array}$$

where Y is FB. Since A' is the same as B' and thus (F A') is the same as (F B)', our induction hypotheses are satisfied.

Now assume $\mathcal{T}^n$ has been constructed. If $\mathcal{P}^{\dagger}$ has only n λ-subpaths we are done. Otherwise we consider cases depending on how the first formula in the n+1-st λ-subpath of $\mathcal{P}^{\dagger}$ was derived (it cannot be by a λ-rule):

a) Let the first formula Y in the $n+1$-st λ-subpath of $\mathcal{P}^{\dagger}$ be derived from the last formula W of the j-th λ-subpath of $\mathcal{P}^{\dagger}$ by an α rule (so Y is W_i, $i = 1$ or 2), and let Y be the successor of the last formula in the k-th λ-subpath of $\mathcal{P}^{\dagger}$. Then W^* is also an α (since W' and $W^{*\prime}$ are the same and neither are λ formulas). Thus it is meaningful to speak of W^*_i, and in fact to form $\mathcal{Q}^{n+1}$ we append

$$W^*_i$$
$$W^*_i{}^{\lambda}$$
$$\vdots$$
$$W^*_i{}^{\underline{\lambda}}$$

below the k-th λ-subpath of $\mathcal{Q}^{n}$. Graphically we have:

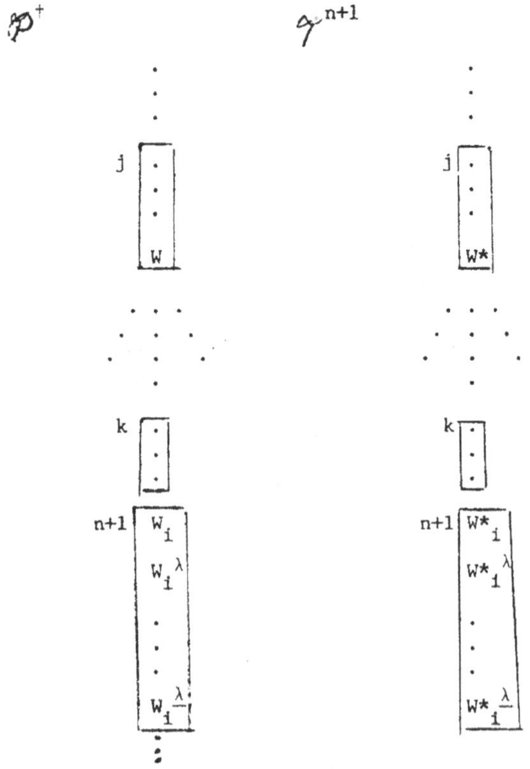

Clearly W_i' is the same as $W^*_i{}'$ and it easily follows that for any member Z of the n+1-st λ-subpath of $\mathcal{P}^\dagger$ and any member Z^* of the n+1-st λ-subpath of $\mathcal{T}^{n+1}$, $Z' = W_i' = W^*_i{}' = Z^{*\prime}$.

b) The case when the first formula of the n+1-st λ-subpath of $\mathcal{P}^\dagger$ is obtained by a β, γ, or δ rule is similar to case a).

c) Let the first formula in the n+1-st λ-subpath of $\mathcal{P}^\dagger$ be a left-hand (right-hand) cut formula W appended below the i-th λ-subpath of $\mathcal{P}^\dagger$. In this case we append a copy of the same λ-subpath on the left (on the right) below the i-th λ-subpath in $\mathcal{T}^n$ to form $\mathcal{T}^{n+1}$. The induction hypothesis is trivially satisfied in this case.

It is easy to see that when we are finished all formulas in $\mathcal{T}$ are obtainable by legal A_1^2 (A_5^2) rules. In particular, all the formulas except the first of any λ-subpath are obtainable by λ rules, and if the first formula of the n-th λ-subpath in is obtained from the last formula in the i-th λ-subpath in $\mathcal{P}^\dagger$ by a certain rule, the first formula of the n-th λ-subpath in $\mathcal{T}$ is obtained from the last formula in the i-th λ-subpath of $\mathcal{T}$ by the same rule. Moreover, since a λ-subpath ending in an atomic formula in $\mathcal{P}^\dagger$ must correspond to a λ-subpath in $\mathcal{T}$ containing the <u>same</u> formula, all the branches in $\mathcal{T}$ must be closed. Thus $\mathcal{T}$ is a proof of B.

Now let us prove:

<u>Theorem 19</u>. Let A be a C^2 formula and let B be any A^2 formula for which B' is A. We can transform any proof $\mathcal{P}$ of A

in C_1^2 into a proof $\mathcal{T}$ of B in A_1^2.

Proof. Let us be given an A and B and $\mathcal{P}$ satisfying the hypotheses of the theorem. By theorem 18 it suffices to find a proof $\mathcal{T}$ of A in A_1^2 (since A' and B' are the same, namely, A).

We first transform $\mathcal{P}$ by removing every axiom sequence. We then replace every occurrence of a parameter P^n that was introduced through an axiom sequence

$$P^n: \quad C(a_1,\ldots,a_i,x_1,\ldots,x_n)$$

by

$$\lambda x_1 \ldots x_n C(a_1,\ldots,a_i,x_1,\ldots,x_n).$$

Now in any part of the tree that came from an expansion tree we remove all but the final formula, which is then obtainable from its predecessor by a λ rule in A_1^2. It is easy to see that the resulting tree is an A_1^2 tableau for A. Every branch, moreover, is either closed or contains a formula $T\ \lambda x_1 \ldots x_n C(a_1,\ldots,a_i,x_1,\ldots,x_n)c_1,\ldots,c_n$ and its conjugate.[6] We extend all such open branches to closed branches to form $\mathcal{T}$.

Finally:

Theorem 20. Let A be a C^2 formula and let B be any A^2 formula for which B' is A. We can transform any proof $\mathcal{P}$ of A in C_5^2 into a proof $\mathcal{T}$ of B in A_5^2.

[6]If we had restricted our C_1^2 tableaux to those that close due to atomic formulas whose predicate parameters were not introduced by axioms, (cf. proposition 30), this case would not arise.

Proof. Let us be given an A, B, and $\mathcal{P}$ satisfying the hypotheses of the theorem. Again by theorem 18 it suffices to find a proof $\mathcal{T}$ of X in A_5^2.

We first remove all axioms

$T(\forall z_1 \ldots z_i)(\exists Z)(\forall x_1,\ldots,x_n)(Zx_1,\ldots,x_n \leftrightarrow C(z_1,\ldots,z_i,x_1,\ldots,x_n))$

and those of their descendents that are the result of γ or δ rule applications to substitute parameters for the $z_1, \ldots, z_i$, Z. Then if

$T(\forall x_1 \ldots x_n)(Px_1,\ldots,x_n \leftrightarrow C(c_1,\ldots,c_i,x_1,\ldots,x_n))$

is a descendent of a (now removed) axiom, we replace it by a cut to

$T(\forall x_1 \ldots x_n)(\lambda x_1 \ldots x_n C(c_1,\ldots,c_i,x_1,\ldots,x_n)x_1,\ldots,x_n \leftrightarrow C(c_1,\ldots,c_i,x_1,\ldots,x_n))$

and its conjugate. **Here we put the part of the tableau that was under the original formula under the T-signed formula, and we extend the** branch containing the F-signed formula to a closed branch, which can easily be seen to be possible. We then substitute $\lambda x_1 \ldots x_n C(c_1,\ldots,c_i, x_1,\ldots,x_n)$ for P throughout the tableau. Clearly C_5^2 rules go into A_5^2 rules under these changes. A closed branch containing a $T\ Pa_1,\ldots,a_n$ and its conjugate may go into one containing $T\lambda x_1 \ldots x_n C(c_1,\ldots,c_i,x_1,\ldots,x_n)a_1,\ldots,a_n$ and its conjugate. In such a case we extend the branch in question to a closed branch as we have noted is always possible. Thus we have found an A_5^2 proof of A.

We note that to obtain an A_5^2 proof of A from a C_5^2 proof we have had to introduce new cuts to take care of the axioms. Thus a cut-free C_5^2 proof may not lead to a cut-free A_5^2 proof.

We now know by Theorems 16, 17, 19 and 20 that A_1^2 is equivalent to C_1^2 and that A_5^2 is equivalent to C_5^2. Since the Hauptsatz that has been proved by Prawitz, Tait, and Takahashi is just to the effect that anything provable in A_5^2 is provable in A_1^2, we can obtain the same result by showing that anything provable in C_5^2 is provable in C_1^2. We will prove this and related theorems in §3. Unfortunately, however, we now have to take up a classical approach, for further constructive proofs have not been obtained.

§2. Second Order Models and Truth Sets

An interpretation $\mathcal{J}$ of C^2 will consist of a sequence $\{U^i\}$ of nonempty sets and a mapping ϕ on the parameters of C^2 such that:

a) For every $n > 0$, each element of U^n is an n-ary relation over U^0, i.e., a subset of $\underbrace{U^0 x \ldots x U^0}_{n}$.

b) For any parameter c^n $(n \geq 0)$, $\phi(c^n) \in U^n$.

We define $\mathcal{J}$-formulas just like C^2 formulas except that parameters c^n are replaced by elements of U^n. We define the truth value of an unsigned (closed) $\mathcal{J}$-formula A as follows:

1) If A is $R^n c_1, \ldots, c_n$, an atomic formula (here $n > 0$, $R^n \in U^n$, $c_i \in U^0$, $i = 1, \ldots, n$), then A is true if $< c_1, \ldots, c_n > \in R^n$ and is false otherwise.

2) If A is $\neg B$, A is true if B is false, and is false otherwise.

3) If A is $(B \wedge C)$, A is true if B is true and C is true, and is false otherwise.

4) If A is $(B \vee C)$, A is true if B is true or C is true, and is false otherwise.

5) If A is $(B \supset C)$, A is false if B is true and C is false, and is true otherwise.

6) If A is $(\exists z^n)B(z^n)$ and there is some $u^n \in U^n$ such that $B(u^n)$ is true, A is true; otherwise A is false.

7) If A is $(\forall z^n)B(z^n)$ and for every u^n in U^n, $B(u^n)$ is true, then A is true; otherwise A is false.

A signed $\mathcal{I}$-formula $T\ A$ is true (false) iff A is true (false). A signed $\mathcal{I}$-formula $F\ A$ is true (false) iff A is false (true).

We now define the truth value under the interpretation $\mathcal{I}$ of a signed or unsigned C^2 formula $X(c_1,\ldots,c_n)$, all of whose parameters are indicated, to be the truth value of

$$X(\phi(c_1),\ldots,\phi(c_n)).$$

Let us denote the above formula also by $X(c_1,\ldots,c_n)^{\phi}$.

We say an interpretation of C^2 in which all the axioms are true is a <u>model</u> of C^2.[7] A model in which each U^n is the power set of $\underbrace{U^0 x \ldots x U^0}_{n}$ (i.e., the set of <u>all</u> n-ary relations over U^0) is called a <u>standard model</u>. A model in which U^0 consists of precisely the parameters of C^2 and where moreover $\phi(c^0) = c^0$ for every parameter c^0 of U^0, is called a <u>parameter model</u>. A model in which for every $n \geq 0$, for each u^n in U^n there is a parameter c^n such that $\phi(c^n) = u^n$ is called an <u>efficient model</u>. We say a model is <u>completely denumerable</u> if each U^n, $n \geq 0$, is denumerable. Clearly every efficient model is completely denumerable.

---

[7] We will assume the interpretations and models discussed in §2 and §3 are for C^2.

We say a set (possibly infinite) of signed or unsigned formulas is satisfiable if there is some model in which every formula of of the set is true. We say a set of signed or unsigned formulas is valid if every formula of the set is true in every model. Let $\overline{S}$, the conjugate of S, be the set consisting of the conjugates[8] of the elements of S. It is easy to see that if a set S is valid, $\overline{S}$ is not satisfiable. In the case of a single formula X (by which we actually mean the unit set of X), X is valid iff $\overline{X}$ is not satisfiable.

We define a second order truth set to be a set E of signed C^2 formulas satisfying the following properties:

1) E contains all the axioms;

2) E contains every C^2 formula or its conjugate, but not both;

3) If $\alpha \in E$, then $\alpha_1 \in E$ and $\alpha_2 \in E$;

4) If $\beta \in E$, then $\beta_1 \in E$ or $\beta_2 \in E$;

5) If $\gamma^n \in E$, then $\gamma^n(c^n) \in E$ for all n-ary parameters c^n.

6) If $\delta^n \in E$, then $\delta^n(c^n) \in E$ for some n-ary parameter c^n.

Throughout §2 and §3 we will use "truth set" to mean "second order truth set."

It is easy to show that in any second order truth set, the converses of conditions 3) - 6), let us call them properties 3') - 6'), also hold. Moreover, any set satisfying properties 1), 2), 3') - 6') is a truth set. It can also be verified without difficulty that:

---

[8] For an unsigned formula A, we may let $\neg A$ be the conjugate of A here.

Proposition 32. The set of all formulas true in an efficient model is a truth set.

The relation between truth sets and models is much closer, however. We first show:

Theorem 21. If E is a truth set, there exists an efficient parameter model for which E is the set of formulas true in the model.

Proof. Let E be a truth set. We let U^0 be the set of parameters in C^2. For every n-ary predicate parameter P^n let

$$R_{P^n} = \{< c_1,\ldots,c_n > \mid T\, P^n c_1,\ldots,c_n \in E\}.$$

Then we let U^n be the set of all the R_{P^n}. We let $\phi(c^0) = c^0$ for every individual parameter, and let $\phi(P^n) = R_{P^n}$ for any n-ary predicate parameter P^n. We have thus defined an interpretation. It can be shown by an induction on degree that every formula of E is true in this interpretation. Since E contains all the axioms, our interpretation is a model and no formula X outside E could be true in our model, for $\overline{X}$ is in E and thus true. Thus E is precisely the set of formulas true in the model. Clearly this model is an efficient parameter model.

Now it is easy enough to construct a model for which the set of true C^2 formulas is not a truth set (see the comment below). However, we can show with a bit of work that:

Theorem 22 . The set of all formulas true in a model $\mathfrak{M}$ of C^2 is a subset of a truth set over a formulation $C^{2'}$ of second order logic that has at most denumerably many parameters not in C^2.

Proof. We have proven a stronger statement for an efficient model. That statement is not true for an arbitrary model because, for instance, if $T(\exists x)A(x)$ is true in a model with sets U^i and map ϕ, this means there is some $u \in U^0$ for which $TA^{\phi}(u)$ is true. But it is sometimes the case that u is not the image of any parameter under ϕ, so that we cannot conclude that $T\,A(a)$ is true in the model for some parameter a. What we will do is show how to obtain from an arbitrary model $\mathfrak{M}$ an efficient parameter model $\mathfrak{M}^P$ of a formulation $C^{2'}$ of second order logic with at most denumerably many new parameters, where the set of true formulas in $\mathfrak{M}^P$, which then must be a truth set by proposition 32, includes the C^2 formulas true in $\mathfrak{M}$. So let us be given a model $\mathfrak{M}$ with sets U^i, $i = 0, 1, \ldots,$ and a parameter map ϕ. For convenience we assume that none of the elements of U^n are in the set $A^n = \{c_1^n, c_2^n, \ldots,\}$ of n-ary parameters of C^2.

We first form a new model $\mathfrak{M}^+$ (with sets U^{i+} and a map ϕ^+) such that the ϕ^+ map is one-to-one on the set $A^0 = \{a_1, a_2, \ldots,\}$ of individual parameters. We let U^{0+} be the union of A^0 and U^0, and define $\phi^+(a_i) = a_i$. For each $R^n \in U^n$, we expand R^n to a new relation R^{n+} over U^{0+} as follows: $\langle e_1, \ldots, e_n \rangle \in R^{n+}$ iff there exists a $\langle u_1, \ldots, u_n \rangle \in R^n$ such that for $i = 1, \ldots, n$, either $e_i = u_i$ or for some a_j, $e_i = a_j$ and $u_i = \phi(a_j)$. We then let U^{n+} be the set of all the R^{n+} and then set $\phi^+(P^n) = [\phi(P^n)]^+$.

It can be shown by an induction on degree that the $\mathfrak{M}$-formula $A(u_1,\ldots,u_n, R_1,\ldots,R_m)$ has the same truth value in $\mathfrak{M}$ as $A(e_1,\ldots,e_n, R_1^+,\ldots,R_m^+)$ has in $\mathfrak{M}^+$, where, for $i = 1, \ldots, n$, either $e_i = u_i$ or for some a_j, $e_i = a_j$ and $\phi(a_j) = u_i$. This immediately implies that if a C^2 formula is true in $\mathfrak{M}$, it is true in $\mathfrak{M}^+$ (i.e., the two models have the same set of true formulas).[9]

From $\mathfrak{M}^+$ we now want to obtain a completely denumerable model with the same set of true formulas. We proceed as follows: let $S_0^0 = A^0$, $S_0^n = \{\phi^+(P^n)\}$ for some $n > 0$. Then assume S_i^n, $n = 1,\ldots,$ have been defined. We now wish to define the S_{i+1}^n, $n = 1,\ldots.$ Let us enumerate the δ^n $\mathfrak{M}^+$-formulas with "parameters" from all the S_i^n: $\delta_1^{n,i}$, $\delta_2^{n,i}$,.... Now if $\delta_j^{n,i}$ is true in $\mathfrak{M}^+$, let $e_j^{n,i}$ be a member of U^{n+} for which $\phi^+[\delta_j^{n,i}](e_j^{n,i})$ is true and if $\delta_j^{n,i}$ is false let $e_j^{n,i}$ be an arbitrary member of U^{n+}. We then define:

$$S_{i+1}^n = S_i^n \cup \{e_j^{n,i}\}_{j=1}^{\infty}.$$

Finally we define:

$$S^n = \bigcup_{i=1}^{\infty} S_i^n.$$

It is clear that each S^n has only denumerably many members. Now if R^{n+} is an element of S^n for $n > 0$, we let R^{n-} be the relation that results when each element $\langle e_1,\ldots,e_n\rangle$ of R^{n+} that contains some component e_i in $U^{0+} - S^0$ is deleted from R^{n+}. Now we define a model $\mathfrak{M}^-$ as follows: We let U^{0-} be S^0 and let U^{n-} be the set of R^{n-} such that $R^n \in S^n$; then we let

[9] Actually it is only at this point that we know the interpretation $\mathfrak{M}^+$ is a model.

$\phi^-(a_i) = \phi^+(a_i) = a_i$; and define $\phi^-(P^n) = [\phi^+(P^n)]^-$. Now let $A(s_1,\dots,s_n, R^1,\dots,R^m)$ be an $\mathcal{M}^+$ formula true (false) in $\mathcal{M}^+$, where $s_i \in S^0$, $i = 1,\dots,n$ and $R^i \in S^{j_i}$, $j_i > 0$, $i = 1,\dots,m$. Let us show (by induction on degree) that $A(s_1,\dots,s_n, R^{1-},\dots,R^{m-})$ is true (false, respectively) in $\mathcal{M}^-$. We give the less obvious steps:

a. If $Rs_1,\dots,s_n$ is true in $\mathcal{M}^+$, then $\langle s_1,\dots,s_n\rangle \in R$, and $\langle s_1,\dots,s_n\rangle \in R^-$, so that $R^-s_1,\dots,s_n$ is true in $\mathcal{M}^-$;

b. Let $(\exists z^n)A(s_1,\dots,s_j,R_1,\dots,R_k,z^n)$ be true in $\mathcal{M}^+$. Then all the parameters $s_1,\dots,s_j,R_1,\dots,R_k$ occur in the S_i^ℓ's by some step i. Thus by construction there will be an $e \in S^n_{i+1}$ such that $A(s_1,\dots,s_n,R_1,\dots,R_m,e)$ is true in $\mathcal{M}^+$. Since $e \in S^n$, we have by induction 1) if $n = 0$, $A(s_1,\dots,s_n,R_1^-,\dots,R_m^-,e)$ is true in $\mathcal{M}^-$ or 2) if $n \neq 0$, $A(s_1,\dots,s_n,R_1^-,\dots,R_m^-,e^-)$ is true in $\mathcal{M}^-$, and hence $(\exists z^n)A(s_1,\dots,s_n,R_1^-,\dots,R_m^-,z^n)$ is true in $\mathcal{M}^-$.

It follows that any C^2 formula which is true in $\mathcal{M}^+$ is true in $\mathcal{M}^-$, i.e., that $\mathcal{M}^-$ is a model that has the same true formulas as $\mathcal{M}^+$. Clearly $\mathcal{M}^-$ is also a completely denumerable model. This already gives us a Skolem-Löwenheim-type theorem:

<u>Proposition 33</u>. If a set of C^2 formulas is satisfiable, it is satisfiable in a completely denumerable model.

We now conclude the proof of the theorem. We recall that each set U^{n-}, $n \geq 0$, is denumerable. Thus if we let $W^n = \{u^n \in U^{n-} \mid \text{there exists a } c^n \text{ such that } \phi^-(c^n) = u^n\}$, the set $V^n = U^{n-} - W^n$ is denumerable (it may be finite or empty; for $n = 0$, $W^n = A^0$, the set of individual parameters of C^2). Let us extend C^2 to a second order logic $C^{2'}$ by adding the elements of V^n

as new (additional) n-ary parameters for each n (in particular, the set of individual parameters of $C^{2'}$ will be the set U^{0-}). We form a (completely denumerable) efficient parameter model $\mathfrak{M}^p$ for $C^{2'}$ as follows: We let the set U^{n^p} be the same as the set U^{n-}. For any individual parameter a in $C^{2'}$ we let $\phi^p(a) = a$. For an n-ary predicate parameter P^n from C^2, we let $\phi^p(P^n) = \phi^-(P^n)$. For an element R^n of V^n, we let $\phi^p(R^n) = R^n$. It is easy to see that all the C^2 formulas have the same truth values in $\mathfrak{M}^-$ and $\mathfrak{M}^p$. Since $\mathfrak{M}^p$ is an efficient model, we obtain the desired result with the use of proposition 32.

Theorem 21 yields:

Proposition 34. If a set of formulas is valid, it is a subset of every truth set.

We can further obtain:

Proposition 35. If a set S of formulas is a subset of every truth set, S is valid.

Proof. It is clear that it suffices to prove that if a single formula X is in every truth set it is valid. So let X be an arbitrary formula. Let $\{b_i^n\}_{i=1}^{\infty}$ be the n-ary C^2 parameters that do not occur in X. We wish to show that for an arbitrary model $\mathfrak{M}$ of C^2, X is true in $\mathfrak{M}$. By theorem 22, the set of formulas true in $\mathfrak{M}$ is a subset of a truth set T' over a denumerable set of parameters that includes all those of C^2. Let $\{d_i^n\}_{i=1}^{\infty}$ be an enumeration of those n-ary parameters in T' that are not found in the formula X. Let Ψ be a mapping on the parameters such that $\Psi(d_i^n) = b_i^n$ and $\Psi(c) = c$ if c occurs in X. Then let T be the set of all formulas

$$Y(\Psi(e_1),\ldots,\Psi(e_n))$$

such that $Y(e_1,\ldots,e_n)$ is a formula in T'. It is not difficult to see that the resulting set T is a C^2 truth set and thus contains X. Thus T' contains X (for X cannot come from any formula of T' other than X). Now it is clear that if the C^2 formulas true in $\mathfrak{M}$ are a subset of T', they must be precisely those formulas of T' that contain only parameters of C^2. So X occurs among the true formulas of $\mathfrak{M}$.

Similarly to proposition 35 we can prove:

<u>Proposition 36</u>. Let S be a set of formulas missing at least denumerably many n-ary parameters of C^2 for every $n \geq 0$. Then if S is satisfiable, it is a subset of some truth set.[10]

Proof. Let S be a set of formulas missing the denumerable set $\{b_i^n\}_{i=1}^{\infty}$ of the n-ary parameters of C^2, and let all the formulas of S be true in a certain model $\mathfrak{M}$. We must show S is a subset of some C^2 truth set. By theorem 22 the set of formulas true in $\mathfrak{M}$ is a subset of a truth set T' over a denumerable set of parameters that includes all those of C^2. Let $\{d_i^n\}_{i=1}^{\infty}$ be an enumeration of those n-ary parameters in T' that are not found in any formula of S (this set must be denumerable because it contains all the elements of $\{b_i^n\}_{i=1}^{\infty}$). Let Ψ be a mapping on the parameters such that $\Psi(d_i^n) = b_i^n$ and $\Psi(c) = c$ if c occurs in some formula of S. Then let T be the set of all formulas

$$Y(\Psi(e_1),\ldots,\Psi(e_n))$$

such that $Y(e_1,\ldots,e_n)$ is a formula in T'. We see again that T

---

[10] That S cannot be an arbitrary set is shown by the example $S = \{F\forall xPx, TPa_1, TPa_2, \ldots\}$, where $a_1, a_2,$ are all the individual parameters of C^2.

is a C^2 truth set. Since S was a subset of T' (S is a subset of the set of all C^2 formulas true in $\mathfrak{M}$, which is a subset of T'), and S goes into itself under the Ψ map, S is a subset of T.

Theorem 21 again easily gives:

Proposition 37. If a set of formulas is a subset of a truth set, it is satisfiable.

We have as corollaries:

Proposition 38. A finite set of formulas is valid iff it is a subset of every truth set.

Proposition 39. A finite set of formulas is satisfiable iff it is a subset of some truth set.

Proposition 40. An unsigned formula A is valid iff T A is a member of every truth set.

Proposition 41. An unsigned formula A is satisfiable iff T A is a member of some truth set.

§3. Consistency and Completeness Proofs

As we have mentioned, this paragraph will present a number of theorems related to the Hauptsatz for the abstraction term formulation of second order logic, which we have seen is the same as the equivalence of C_1^2 and C_5^2. We will discuss a number of systems $C_0^2, C_1^2, \ldots, C_5^2$, where any tableau in $C_0^2, \ldots, C_4^2$ will be a tableau in C_5^2. When speaking of arbitrary tableaux we will include tableaux for infinite sets S of formulas. In such a case the members of S are allowed to be introduced into a tableau at any step (as with axioms in C_5^2). When speaking of closed tableaux, however, we may assume the elements of S needed are at the top, for we can always transform an arbitrary closed tableau into a finite one satisfying the property in question.

One can show in a manner almost identical to that used in first order logic (cf. Smullyan [1], p. 55):

Theorem 23. a) If S is an arbitrary satisfiable set of C^2 formulas, no C_5^2 (and hence no $C_0^2,\ldots,C_4^2$) tableau for S can close.
b) If P is provable in C_5^2 (or $C_0^2,\ldots,C_4^2$), then P is valid.

We say this theorem expresses the semantic consistency of $C_0^2,\ldots,C_5^2$.

Now we would like to prove the equivalence of C_1^2 and C_5^2, (i.e., obtain the Hauptsatz for the abstraction term formulation). We can do this (as in first order logic) by showing that everything valid is provable in C_1^2 (and hence in C_5^2), in which case the set of formulas provable in either system will be precisely the set of valid formulas. Before doing this, however, we will introduce an apparently less restricted system C_0^2:

The System C_0^2 is the same as the system C_1^2 except that an axiom sequence

$$P^n:\ A$$

is not required to be followed by a formula $\gamma^n(P^n)$ resulting from a γ^n rule application.

We can easily show constructively, however:

Proposition 42. The systems C_0^2 and C_1^2 are equivalent.

Proof. Let an arbitrary C_0^2 proof $\mathcal{P}$ be given. We can assume the proof $\mathcal{P}$ is finite so that only finitely many predicate parameters $P_1,\ldots,P_n$ are introduced by axiom sequences $P_i: A_i$, $i = 1,\ldots,n$. We can also choose n denumerable sequences of distinct parameters

$$Q_{i_1},\ldots,Q_{i_j},\ldots, \quad i = 1,\ldots,n$$

such that if P_i $(i = 1,\ldots,n)$ is an m-ary predicate parameter all the Q_{i_j} are m-ary predicate parameters that do not occur in the proof. We transform $\mathcal{P}$ to a closed C_1^2 proof of the same formula as follows:

1) For each $i = 1,\ldots,n$ we enumerate the occurrences of a $\gamma(P_i)$ in the proof (actually there can only be finitely many such occurrences) and then replace the j-th occurrence $\gamma_j(P_i)$ by:

$$Q_{i_j}: A_i$$
$$\gamma_j(Q_{i_j}).$$

We then replace all the occurrences of P_i in the proof that came from its introduction in $\gamma_j(P_i)$ by Q_{i_j}.

2) We eliminate the occurrences of P_i: A_i.

3) For any open branch containing a $\pi Q_{i_j} c_1,\ldots,c_m$ and $\bar{\pi} Q_{i_k} c_1,\ldots,c_m$ that come from a $\pi P_i c_1,\ldots,c_m$ and $\bar{\pi} P_i c_1,\ldots,c_m$, we insert the appropriate expansion subtrees after the formulas in question, which yields as the only potentially open branch one containing a formula and its conjugate. We then extend this branch to an atomically closed branch with the use of α, β, γ, and δ rules.

We thus see that the <u>important</u> property of C_1^2 can be expressed approximately as follows: although we are allowed to assume - where the a's are arbitrary and the b's individual parameters - that for any statement $A(a_1,\ldots,a_i, b_1,\ldots,b_n)$ there is a predicate P^n such that $P^n c_1,\ldots,c_n$ is true iff $A(a_1,\ldots,a_i,c_1,\ldots,c_n)$ is, we can only explore this fact in terms of some P^n when our <u>original</u> assumptions lead us to conclude $\pi P^n c_1,\ldots,c_n$ (which must be through a conclusion referring to <u>all</u> n-ary predicates P.)

We now prove the completeness of C_0^2, and thus by proposition 42, of C_1^2.[11] We will use the concept of a <u>Hintikka set</u> (related to Hintikka's model sets [1], and Schütte's semi-valuations [6]) <u>for</u> C_0^2:

Let H be a set of (closed) signed formulas and let each (open) C^2 formula $A(x_1,\ldots,x_n)$[12] have an n-ary predicate parameter P^n associated with it. Then H is a C_0^2 <u>Hintikka set</u> (with respect to the association mentioned) if the following conditions are satisfied:

1) For no formula X are both X and $\overline{X}$ in H.

2) If $\alpha \in H$, then $\alpha_1 \in H$ and $\alpha_2 \in H$.

3) If $\beta \in H$, then $\beta_1 \in H$ or $\beta_2 \in H$.

4) If $\gamma^n \in H$, then $\gamma^n(c^n) \in H$ for each n-ary parameter c^n $(n \geq 0)$.

5) If $\delta^n \in H$, then $\delta^n(c^n) \in H$ for some n-ary parameter c^n $(n \geq 0)$.

6) If $\pi P^n c_1,\ldots,c_n \in H$ and there is a formula $A(x_1,\ldots,x_n)$ with which P^n is associated, then $\pi A(c_1,\ldots,c_n)$ is in H.

Half of the completeness proof is embodied in:

---

11 This proof corresponds to the Prawitz and Takahashi proofs of the Haupsatz for type theory.

12 Here all the free variables of the formula are denoted, although all variables denoted need not occur in the formula. Note that the free variables are all individual variables.

Proposition 43. Every C_0^2 Hintikka set is satisfiable.

Proof. Let H be a C_0^2 Hintikka set with respect to a given association between formulas and (certain) predicate parameters. We form an interpretation $\mathcal{Q}$ for the (closed) C_0^2 formulas as follows. We let U^0 be the set of individual parameters of C_0^2. Now let P^n be an arbitrary n-ary predicate parameter of C_0^2. We say a relation R^n over the individual parameters is a possible value of P^n with respect to H if:

a) $T\ P^n c_1,\ldots,c_n \in H$ implies $\langle c_1,\ldots,c_n\rangle \in R^n$;

b) $F\ P^n c_1,\ldots,c_n \in H$ implies $\langle c_1,\ldots,c_n\rangle \notin R^n$.

Then we let

$$U^n = \{R^n \mid \text{there is some } P^n \text{ such that } R^n \text{ is a possible value of } P^n\}.$$

We note that although U^0 is the set of parameters, our interpretation $\mathcal{Q}$ is certainly not necessarily completely denumerable. For example, if for some P^n, H does not contain any atomic formula $\pi P^n c_1,\ldots,c_n$, the set of possible values of P^n will be the power set of $\underbrace{U^0 \times \ldots \times U^0}_{n}$, which is not a denumerable set (i.e., U^n will not be denumerable in such a case). Before defining a parameter map ϕ, we recall that the truth value of $\mathcal{Q}$-formulas is already well-defined when the sets U^n have been defined. We note the straightforward

Lemma 1. If $Q_1,\ldots,Q_m$ are all the predicate parameters of a formula $A(Q_1,\ldots,Q_m)$ and for each $i = 1,\ldots,m$, R_{Q_i} is a possible value of Q_i, then:

$$\pi A(Q_1,\ldots,Q_m) \in H \text{ implies } \pi A(R_{Q_1},\ldots,R_{Q_m}) \text{ is true.}$$

Proof. We prove this lemma by induction on the degree of $\pi A(Q_1,\ldots,Q_m)$.

a) Let $\pi A(Q_1,\ldots,Q_m)$ be the atomic formula $T\ Qc_1,\ldots,c_n$, and let R_Q be a possible value of Q. Then if $T\ Qc_1,\ldots,c_n \in H$, $<c_1,\ldots,c_n> \in R_Q$ since R_Q is a possible value of Q. Thus $T\ R_Q c_1,\ldots,c_n$ is true. The case of an atomic formula $F\ Qc_1,\ldots,c_n$ is similar.

b) Let $\pi A(Q_1,\ldots,Q_m)$ be an α formula in H. Then since H is a C_0^2 Hintikka set $[\pi A(Q_1,\ldots,Q_m)]_i \in H$, $i = 1,2$. By the induction hypothesis, $[\pi A(R_{Q_1},\ldots,R_{Q_m})]_i$ is true for $i = 1,2$. But then $\pi A(R_{Q_1},\ldots,R_{Q_m})$ is true. The case when $\pi A(Q_1,\ldots,Q_m)$ is a β formula is similar.

c) Let $\pi A(Q_1,\ldots,Q_m)$ be a γ^0 formula $\gamma^0(Q_1,\ldots,Q_m)$ that is in H. Then since H is a C_0^2 Hintikka set, for every individual parameter c, $\gamma^0(Q_1,\ldots,Q_m)(c) \in H$. By the induction hypothesis, $\gamma^0(R_{Q_1},\ldots,R_{Q_m})(c)$ is true for every $c \in U^0$. But then $\gamma^0(R_{Q_1},\ldots,R_{Q_m})$ is true. The case when $\pi A(Q_1,\ldots,Q_m)$ is a δ^0 formula is similar.

d) Let $\pi A(Q_1,\ldots,Q_m)$ be a γ^n formula $\gamma^n(Q_1,\ldots,Q_m)$, $n > 0$, which is in H. Then since H is a C_0^2 Hintikka set, $\gamma^n(Q_1,\ldots,Q_m)(Q^n)$ is in H for every n-ary predicate parameter Q^n. Then if R^n is any possible value of any Q^n (i.e., if R^n is <u>any</u> element of U^n), by the induction hypothesis

$\gamma^n(R_{Q_1},\dots,R_{Q_m})(R^n)$ is true. But then $\gamma^n(R_{Q_1},\dots,R_{Q_m})$ is true.

e) Let $\pi A(Q_1,\dots,Q_m)$ be a δ^n formula $\delta^n(Q_1,\dots,Q_m)$, $n > 0$, which is in H. Then since H is a C_0^2 Hintikka set, $\delta^n(Q_1,\dots,Q_m)(Q^n)$ is in H for some n-ary predicate parameter Q^n. Then by the induction hypothesis, if R_{Q^n} is a possible value of Q^n (which is thus a member of U^n), $\delta^n(R_{Q_1},\dots,R_{Q_m})(R_{Q^n})$ is true. But then $\delta^n(R_{Q_1},\dots,R_{Q_m})$ is true.

We note that for the proof of this lemma we have not needed to use the fact that <u>every</u> possible value of every n-ary predicate parameter P^n is in U^n, (i.e., we could have put only one possible value for each P^n into U^n and still have obtained lemma 1) We will need that stronger fact, however, to obtain

<u>Lemma 2</u>. All the axioms are true (and thus for any ϕ chosen, $\mathcal{A}$ will be a model).[13]

Proof. Let Z be the axiom

$T(\forall z_1 \dots z_i)(\exists X)(\forall x_1 \dots x_n)(Xx_1,\dots,x_n \leftrightarrow A(z_1,\dots,z_i,x_1,\dots,x_n))$

We must be able to show that

(*) $(\exists X)(\forall x_1 \dots x_n)(Xx_1,\dots,x_n \leftrightarrow A(d_1,\dots,d_i,x_1,\dots,x_n))$ is true

---

[13]Since the axioms have no parameters they are true as $\mathcal{J}$-formulas iff they are true in the interpretation (where the map ϕ is taken into account) as C^2 formulas.

for any $d_1,\ldots,d_i$ such that if z_j is an n_j-ary variable, $d_j \in U^{n_j}$, $j = 1,\ldots,i$. Let us rewrite (*) as

(**) $(\exists X)(\forall x_1 \ldots x_n)(Xx_1,\ldots,x_n \leftrightarrow B(e_1,\ldots,e_j,R_1,\ldots,R_k,x_1,\ldots,x_n))$,

where we distinguish between the d's that are individuals: $e_1,\ldots,e_j$; and the d's that are predicates: $R_1,\ldots,R_k$. Now let $Q_1,\ldots,Q_k$ be arbitrary predicate parameters for which $R_1,\ldots,R_k$ are possible values, and let P^n be the predicate parameter that is associated with

$$B(e_1,\ldots,e_j,Q_1,\ldots,Q_k,x_1,\ldots,x_n)$$

as a part of the assumption about H. We will see that the set R^n of $\langle c_1,\ldots,c_n\rangle$ such that

$$B(e_1,\ldots,e_j,R_1,\ldots,R_k\ ,c_1,\ldots,c_n)$$

is true is a possible value of P^n. For if $T\ P^n c_1,\ldots,c_n \in H$, then by property 6 of a C_0^2 Hintikka set, $T\ B(e_1,\ldots,e_j,Q_1,\ldots,Q_k\ ,c_1,\ldots,c_n)$ is also in H. But then by lemma 1, $B(e_1,\ldots,e_j,R_1,\ldots,R_k\ ,c_1,\ldots,c_n)$ is true, i.e., $\langle c_1,\ldots,c_n\rangle \in R^n$. And the case for $F\ P^n c_1,\ldots,c_n$ in H is analogous. Now, since R^n is a possible value of P^n, and U^n contains <u>all</u> the possible values of P^n, $R^n \in U^n$ and is clearly a member of U^n that "makes (**) true," i.e., obviously for every $\langle c_1,\ldots,c_n\rangle$,

$$R^n c_1,\ldots,c_n \leftrightarrow B(e_1,\ldots,e_j,R_1,\ldots,R_k,c_1,\ldots,c_n)$$

is true.

This completes the proof of lemma 2. Now we can quickly finish the proof of proposition 43.

We finally choose a ϕ such that $\phi(c) = c$ for an individual parameter c, and $\phi(P^n)$ is a possible value of P^n for any predicate parameter P^n. By lemma 2, $\mathcal{J}$ is a model. And by Lemma 1, every element of H is true in the model. Thus we have shown that H is satisfiable and we are done.

Now we prove:

Theorem 24. Every valid (unsigned) formula is provable in C_0^2.

Proof. Let A be a formula that is not provable in C_0^2, i.e., no C_0^2 tableau for F A is closed. We must show that A is not valid, i.e., that F A is satisfiable. We will do this by giving a procedure for obtaining a so-called complete C_0^2 tableau $\mathcal{T}_X$ for any signed formula X, and showing that if the complete tableau for a formula X is not closed, X is satisfiable. And we will do this by showing that the set of formulas on any open branch of a complete C_0^2 tableau for a formula X contains a C_0^2 Hintikka set (of which the formula X is a member).

So let X be an arbitrary signed formula. Let $A_1, \ldots, A_i, \ldots,$ be a single enumeration of all the unsigned formulas $A(x_1, \ldots, x_n)$ (for all integers n) where $A(x_1, \ldots, x_n)$ has at most n free individual variables and no free predicate variables.[14] The complete C_0^2 tableau for X is the tableau $\mathcal{T}_X$ constructed as follows.

At step 1 we make X the origin of a tree. Then if A_1 is $B(x_1, \ldots, x_m)$, we choose some m-ary predicate parameter P^m not in X or A_1 and adjoin the axiom sequence

$$P^m\colon\ A(x_1, \ldots, x_m)$$

below X. We declare all the formulas of the axiom sequence to be used.

[14] One might better say this is a listing of pairs, one member of which is an open formula, the other of which is a nonempty set of individual variables including all the variables of the formula.

Now assume steps $1,\ldots,n$ have been completed and that certain formulas in $\mathcal{Y}$ have been declared to be used at earlier steps. At step $n+1$, we proceed as follows: First we let $\theta_1,\ldots,\theta_i$ be all the branches of the tree. If A_{n+1} is $B(x_1,\ldots,x_m)$ we take distinct m-ary predicate parameters $P_1,\ldots,P_i$ not occurring in the tree or A_{n+1} and extend θ_j, $j = 1,\ldots,i$ by adjoining the axiom sequence

$$P_j: \ A_{n+1}$$

below its end point. We declare all the points of these axiom sequences used. Secondly, we let Y be the left-most unused point in the tree to which some rule is applicable (if such a point exists; otherwise we are done with step $n+1$). We then extend all the branches $\theta'_1,\ldots,\theta'_j$ through Y according to the C_0^2 rule that is applicable to Y:

a) If Y is an α we adjoin Y_1 and Y_2 below the end point of each θ'_k, $k = 1,\ldots,j$.

b) If Y is a β, we branch to Y_1 and Y_2 below the end-point of each θ'_k, $k = 1,\ldots,j$.

c) If Y is a γ^n, $n \geq 0$, we adjoin $Y(c^n)$, Y^n below the end point of each θ'_k, $k = 1,\ldots,j$,[15] where c^n is the first parameter in the list of the n-ary parameters such that $Y(c^n)$ does not occur in θ'_k. If there is no such parameter for a branch θ'_k, we do not extend θ'_k.

d) If Y is a δ^n, $n \geq 0$, we choose j n-ary parameters $c_1,\ldots,c_j$ which do not occur in the tree and extend θ'_k, $k = 1,\ldots,j$

---

15 Here we say we used the repetition rule as a derived rule to derive Y, but we will eliminate all repetitions from our tree later.

by adjoining $Y(c_k)$.

e) If Y is $T\ P^n c_1,\ldots,c_n$ ($F\ P^n c_1,\ldots,c_n$), where an axiom sequence

$$P^n: \ A(x_1,\ldots,x_n)$$

occurs above Y, we extend each $\theta'_1,\ldots,\theta'_j$ by adjoining the T- (F- respectively) expansion subtree for P^n: $A(x_1,\ldots,x_n)$ and $<c_1,\ldots,c_n>$ below the end point of the branch.

The tree $\mathcal{T}$ that results from the above procedure may contain some uses of the repetition rule. We let $\mathcal{T}_X$ be the C_0^2 tableau that results when we eliminate these uses, which can clearly be done in a systematic way. Now assume $\mathcal{T}_X$ has an open branch θ. Clearly θ is infinite, because our putting axiom sequences in at every step forced all the branches to be infinite. We wish to show that the set of formulas in θ has as a subset a C_0^2 Hintikka set containing X. Let H be the set of formulas in θ defined as follows:

1) X is in H;

2) if a formula Z is in H, all descendents of Z in θ by α, β, γ^n, or δ^n rules are in H;

3) if an atomic formula $\pi P^n c_1,\ldots,c_n$ is in H, and the axiom sequence

$$P^n: \ A(x_1,\ldots,x_n)$$

is in θ, then the formula $\pi A(c_1,\ldots,c_n)$ - which must be in θ since by the construction of $\mathcal{T}_X$, the π expansion subtree for P^n: $A(x_1,\ldots,x_n)$ and $<c_1,\ldots,c_n>$ must be used at some point in every branch through $\pi P^n c_1,\ldots,c_n$, and only the branch through the final formula $\pi A(c_1,\ldots,c_n)$ can be open- is in H. Now it is easy to see that our set H satisfies the requirements in the definition of a C_0^2 Hintikka set. The association between formulas $A(x_1,\ldots,x_n)$ and predicate parameters that is assumed is the obvious one: the predicate parameter P^n is associated with the formula $A(x_1,\ldots,x_n)$ if

$$P^n: \ A(x_1,\ldots,x_n)$$

occurs in θ (by our construction some such axiom sequence occurs in θ for every formula $A(x_1,\ldots,x_n)$).

As mentioned at the beginning, this result yields the theorem. For if A is not provable, the complete C_0^2 tableau $\mathcal{T}_{F\,A}$ is not closed, i.e. has an open branch θ. Such a branch has as a subset a C_0^2 Hintikka set H containing F A. By proposition 43 this set is satisfiable, so that F A is satisfiable and A is not valid.

Theorems 23 and 24 yield:

Theorem 25. The systems C_0^2 and C_5^2 are equivalent.

Proof. Clearly any formula provable in C_0^2 is provable in C_5^2. If a formula A is provable in C_5^2, it is valid by theorem 23, and thus provable in C_0^2 by theorem 24.

Theorem 25 and Proposition 42 finally give as (our Haupsatz equivalent):

Theorem 26. The systems C_1^2 and C_5^2 are equivalent.

We now recall that the interest of the proof-theorist in theorem 26 is due to the fact that a constructive proof of this theorem (or of theorem 25) would yield a constructive proof of the synthetic consistency of second order logic -- we will review this again at the end of this paragraph. The problem for the proof-theorist is thus to try to find some constructive way of going from a C_5^2 proof to a C_1^2 proof. But the system C_1^2 has a number of restrictions on it compared to C_5^2. One might ask if we couldn't consider systems less restricted than C_5^2 whose equivalence with C_5^2 would still yield synthetic consistency, for in this case a constructive procedure for going from a C_1^2 proof to a proof in the restricted system might be easier to find. We will now consider two such possible systems, which we will call C_2^2 and C_3^2.

The System C_2^2 is like C_1^2 except that restriction P_3 is not made, i.e., the last formula of an axiom sequence can be expanded at any time and need not occur in an expansion subtree. In other words, C_2^2 is like C_5^2 except that:

P_1. The cut rule is not used; and

P_2. The axiom rule is only used with an axiom sequence of the form

$$P^n: \ A(x_1,\ldots,x_n),^{16}$$

and only when the next rule applied is a γ^n rule in which P^n is substituted for a variable.

Of course, since anything provable in C_5^2 is provable in

---

16 Where the assumptions about variables is as usual.

C_1^2, and anything provable in C_1^2 is provable in C_2^2, the proofs of the beginning of this paragraph yield the equivalence of C_2^2 and C_5^2. We will show, however, that we can obtain that equivalence with a (classical) proof that is somewhat shorter and more straightforward.

Theorem 27. Every valid formula is provable in C_2^2.

Proof. Let A be a formula that is not provable in C_2^2, i.e., no C_2^2 tableau for F A is closed. We must show that A is not valid or that F A is satisfiable. As before, we will give a procedure for obtaining a so-called complete C_2^2 tableau $\mathcal{T}_X$ for an arbitrary (signed) formula X, and show that the set of formulas on an open branch of a complete C_2^2 tableau is satisfiable (in this case we will not have to consider any special subset of the set.)

So let X be an arbitrary signed formula. For each n, let $A_1^n,\ldots,A_i^n,\ldots$, be an enumeration of all the (unsigned) formulas $A(x_1,\ldots,x_n)$ with at most n free individual variables and no free predicate variables. The complete C_2^2 tableau for X is the tableau $\mathcal{T}_X$ constructed as follows:

At step 1 we make X the origin of a tree $\mathcal{T}$.

Now assume steps $1,\ldots,n$ have been completed and that certain formulas in $\mathcal{T}$ have been declared used at earlier steps. At step n+1, if $\mathcal{T}$ is closed or all the nonatomic formulas on all its open branches have been used, we stop. Otherwise we let Y be the left-most unused nonatomic point of lowest level that is on some open branch in $\mathcal{T}$. We let $\theta_1,\ldots,\theta_i$ be all the open branches in $\mathcal{T}$ that contain Y. We then extend each of these branches in the following manner:

a) If Y is an α, we adjoin Y_1, Y_2 below the end point of each θ_j, $j = 1,\ldots,i$.

b) If Y is a β, we branch to Y_1 and Y_2 below the end point of each θ_j, $j = 1,\ldots,i$.

c) If Y is a γ^n, we adjoin $Y(c_j^n)$,Y below the end point of each θ_j, where c_j^n is the first n-ary parameter such that $Y(c_j^n)$ does not occur in θ_j - if there is no such parameter for a branch θ_j we do not make the extension. Then if $n > 0$, we make the further extension: We let $A_j(x_1,\ldots,x_n)$ be the first formula in the list $\{A_m^n\}_{m=1}^{\infty}$ such that no axiom sequence

$$P^n:\ A_j(x_1,\ldots,x_n)$$

occurs for $A_j(x_1,\ldots,x_n)$ in θ_j. We choose i distinct n-ary parameters $P_1,\ldots,P_i$ that occur neither in the tree nor in any of the $A_j(x_1,\ldots,x_n)$. We then further extend θ_j $(j = 1,\ldots,i)$ by adjoining

$$P_j:\ A_j(x_1,\ldots,x_n)$$

$$Y(P_j)$$

d) If Y is a δ^n, we choose i distinct n-ary parameters $c_1,\ldots,c_i$, which do not occur in the tree and extend θ_j by adjoining $Y(c_j)$.

The tree $\mathcal{T}$ that results from the above procedure may contain some uses of the repetition rule. We let $\mathcal{T}_X$ be the C_2^2 tableau that results when we eliminate these uses.

We now want to show that if $\mathcal{T}_X$ has an open branch θ, the set of formulas in θ, and thus X, must be satisfiable. So let θ be an open branch of $\mathcal{T}_X$. We will consider two cases in showing θ satisfiable:

<u>Case 1</u>. Let no γ^n formula, $n > 0$, occur in θ.

In this case the model constructed is just the same as in the first order completeness proof (see Smullyan [1], pp. 58-60). Thus let U^0 be the set of all individual parameters and let U^n be the set of <u>all</u> n-ary relations over U^0. We let $\phi(a) = a$ for each individual parameter a, and for any n-ary predicate parameter P let $\phi(P) = \{<c_1,\ldots,c_n> \mid T\ P^n c_1,\ldots,c_n \in \theta\}$

It is clear that every axiom is true in this interpretation, so that we have a model. We see that all the elements Y of θ are true (just as in the first order case) by induction on the degree of Y. Thus if Y is of degree 0 it is immediate that Y is true. Now suppose that Y is of positive degree and that every element of θ of lower degree is true. If Y is an α, both Y_1 and Y_2 are in θ and are true because they are of lower degree than Y. Hence α is true. If Y is a β, either Y_1 or Y_2 is in θ, is of lower degree than Y, and thus is true. Hence Y is true. If Y is a γ^0, then for every individual parameter a, $\gamma^0(a)$ is in θ, is of lower degree than Y, and is thus true. So γ^0 is true. If Y is a δ^n $(n \geq 0)$, then $Y^n(c^n)$ is in θ for some c^n of the appropriate kind, is of lower degree, and is thus true. Hence Y is true.

Thus we see we have a model satisfying all the formulas of θ in Case 1.

Case 2. Let some γ^n, $n > 0$, occur in θ.

The first thing we note is that in this case, for every (closed signed) formula Y, either Y or $\overline{Y}$ occurs in θ. For let Y be an arbitrary formula πB, which we also denote by $\pi C(x_1,\ldots,x_n)$.[17] Since in this case θ contains an axiom sequence for every formula $D(x_1,\ldots,x_n)$, it must contain some axiom sequence

$$P^n: \; C(x_1,\ldots,x_n)$$

whose last formula is

$$T(\forall x_1\ldots x_n)(P^n x_1,\ldots,x_n \leftrightarrow B).$$

Considering successive rule applications we find that θ contains

$$T\ P^n c_1,\ldots,c_n \leftrightarrow B$$

for every n-tuple of individual parameters $\langle c_1,\ldots,c_n\rangle$, and thus also the β formulas

$$T\ P^n c_1,\ldots,c_n \supset B$$

and

$$T\ B \supset P^n c_1,\ldots,c_n.$$

Then θ contains either $F\ P^n c_1,\ldots,c_n$ or $T\ B$ and also either $F\ B$ or $T\ P^n c_1,\ldots,c_n$. But since θ is not closed, it cannot contain both $F\ P^n c_1,\ldots,c_n$ and $T\ P^n c_1,\ldots,c_n$, so it must contain either $T\ B$ or $F\ B$, i.e., Y or $\overline{Y}$.

Also we can see that for every $m > 0$, every m-ary axiom is in θ. For otherwise (by the above) there would be an m-ary axiom

[17] Where clearly $x_1,\ldots,x_n$ are individual variables without actual occurrences in the formula in question. Again we note that we could have restricted ourselves to axioms where the relevant individual variables do occur, but our proof would become slightly more complicated.

whose conjugate

(*) $F(\forall z_1 \dots z_i)(\exists X^m)(\forall x_1 \dots x_m)(X^m x_1,\dots,x_m \leftrightarrow B(z_1,\dots,z_i,x_1,\dots,x_m))$

is in θ. Then for some parameters $c_1,\dots,c_i$,

$$F(\exists X^m)(\forall x_1 \dots x_m)(X^m x_1,\dots,x_m \leftrightarrow B(c_1,\dots,c_i,x_1,\dots,x_m))$$

would be in θ, a γ^m formula. But then again every m-ary axiom would be in θ by our construction.

But now it is clear that the set of formulas in θ form a truth set E. By theorem 21, E is satisfiable in a (completely denumerable) efficient model. (We note that, in Case 1 we have used what might be called a first order parameter model.) Thus in Case 2 also, we have seen that we can find a model satisfying all the formulas of θ. This yields theorem 27.

We can see that the proof of theorem 27 we have given is somewhat simpler than the proof of theorem 26. For instance in the proof of theorem 26 we used the concept of a C_0^2 Hintikka set, which was significantly more complicated than that of a first order Hintikka set. We needed lemma 1 and lemma 2 to show such a set satisfiable and the model we found was not a completely denumerable parameter model. In the proof of theorem 27, on the other hand, Case 1 merely uses a trivial extension of a first order proof. Case 2 was able to use the simple proof of theorem 21 to obtain a completely denumerable parameter model because the complete tableau construction had forced the elements of θ to form a truth set. We note how similar to first order techniques are those used in this proof.

As in theorem 25, now theorems 23 and 27 yield:

Theorem 28. The systems C_2^2 and C_5^2 are equivalent.

When we look at the restrictions in C_2^2 - that cuts may not be used, and that axioms may not be used to introduce a parameter P^n unless P^n is going to be used in a γ^n $(n > 0)$ rule immediately afterwards - we might consider further weakening the second restriction. Thinking of our syntactic goal, we might consider a system in which an axiom sequence is only allowed to be used on a branch to introduce an n-ary predicate parameter when a γ^n formula already occurs on the branch, and in fact, we can prove a completeness theorem for such a system in a (very) slightly simpler manner than that used in the proof of theorem 31. But this requirement now seems quite artificial, lacking in meaning in terms of the semantics of the systems in question. Let us then consider one last system, much less restricted than C_2^2, which has a more obvious semantic meaning.

The System C_3^2 is like the system C_5^2 except for the following two restrictions:

P_1. The cut rule is not allowed to be used.

P_2. The axiom rule is only allowed to be used in a tableau for a set of formulas S when at least one formula in S is not a first order formula.

The equivalence of C_3^2 and C_5^2 is easily obtained:

Theorem 29. The systems C_3^2 and C_5^2 are equivalent.

Proof. Since C_4^2 (C_5^2 without the cut rule) and C_5^2 have been shown equivalent, it suffices to show C_3^2 and C_4^2 equivalent. Clearly anything provable in C_3^2 is provable in C_4^2. Also any C_4^2 proof of a formula that is not a first order formula is also a C_3^2 proof of that formula. We need only see, then, that any first order formula provable in C_4^2 is provable in C_3^2. But if such a formula A is provable in C_4^2, it is valid by theorem 23, i.e., is true in all (second order) models. In particular it is true in all standard models, which is the same thing as saying that it is first order valid. By the first order Gödel completeness theorem (see Smullyan [1], p. 60), it is provable without cut in the first order tableau system, i.e., it is provable in C_3^2.

Theorem 30. As usual we have: The equivalence of any one of the systems $C_0^2,\ldots,C_3^2$ with C_5^2 implies (constructively) the syntactic consistency of C_5^2 (and thus of $C_0^2,\ldots,C_4^2$).

Now let us recall Takeuti's work [1]. From it we can obtain the following result: Let $\{A_1,\ldots,A_n,\ldots\}$ be the number-theoretic axioms one adjoins to second order logic to obtain second order arithmetic. Assume $0 = 1$ can be proven in second order arithmetic. This means that there are finitely many of the axioms $A_{i_1},\ldots,A_{i_m}$ such that $(A_{i_1} \wedge \ldots \wedge A_{i_m}) \supset 0 = 1$ is provable in second order logic (the system Takeuti used was like A_5^2; by our results of §1, we can take second order logic to be C_5^2). Takeuti's work shows constructively how to go from such a proof to a proof (in second order logic) of a formula $(B_1 \wedge \dots \wedge B_k) \supset 0 = 1$, where the B_i are axioms of first order number theory. But if we consider C_1^2, C_2^2, and C_3^2, we see that any proof of a first order formula in one of these systems is a proof in first order logic. Thus by the equivalence of these systems with C_5^2, we obtain, under our original assumption, that $0 = 1$ is provable in first order arithmetic. Since this has been proven (constructively) not to be the case (Gentzen [1], [2], Ackermann [1], Gödel [1]), we obtain that one cannot prove $0 = 1$ in second order arithmetic. The non-constructive link in this argument is, of course, the equivalence of one of the systems C_1^2, C_2^2, or C_3^2 with C_5^2.

§4. Logical Frameworks for Higher Order Logic and Type Theory - The Henkin Completeness Theorem

Smullyan [2] has developed a subject he calls "abstract quantificiation theory," the goal of which is to abstract these key properties of formulas and proofs which are used to obtain the mathematically interesting results of first order logic. It turns out that higher finite-ordered logics and type theory can be interpreted in such frameworks (if we add the concept of axioms). We will show, however, that these logics and quantification theory can all be interpreted within what might be called "logical frameworks for many-sorted theories." The use of these frameworks rather than those described by Smullyan will make our discussion somewhat simpler.

Thus by a (classical) many-sorted logical framework $\mathcal{S}$ we shall mean an ordered eight-tuple $\langle E, A, C, D, \phi, -, I, \{P^\tau\}_{\tau\in I}\rangle$ satisfying the following conditions:

(1) E is a denumerable set whose elements we call the elements of the framework.

(2) A is a subset of E, called the axioms of the framework.

(3) C is a subset of E, whose elements we call conjunctive elements.

(4) D is a subset of E, whose elements we call disjunctive elements.

Elements in $C \cup D$ will be called compound; elements of E outside $C \cup D$ will be called atomic.

(5) ϕ is a function which assigns to every compound element x a finite or denumerable sequence $\langle x_1, x_2, \ldots, x_n, \ldots\rangle$ (of elements of E) whose terms we call the components of x -- more specifically, we refer to x_i as the i-th component of x. We require that if x be both conjunctive and disjunctive, then x has only one component.

(6) - is a function which assigns to each x in E an element $\bar{x}$ in E called the conjugate of x. We require that conjugation obeys the following laws (for all x in E):

(i) $\bar{x} \neq x$;

(ii) $\bar{\bar{x}} = x$;

(iii) If x is conjunctive, $\bar{x}$ is disjunctive, and if x is disjunctive, then $\bar{x}$ is conjunctive. (From this it follows that if x is atomic, so is $\bar{x}$.)

(iv) $\phi(\bar{x})$ has the same number of terms as $\phi(x)$, and the i-th term of $\phi(\bar{x})$ is the conjugate of the i-th term of $\phi(x)$. (Thus, if $\phi(x) = \langle x_1, x_2, \ldots, x_n, \ldots\rangle$ then $\phi(\bar{x}) = \langle \bar{x}_1, \bar{x}_2, \ldots, \bar{x}_n, \ldots\rangle$.)

(7) I is a nonempty set, the members of which we call <u>types</u>.

(8) $\{P^\tau\}_{\tau\epsilon I}$ is a set of functions, one for each type τ in I. If $\tau \epsilon I$, P^τ is a function which assigns to each x in E a finite set $P^\tau(x)$ of positive integers where for any x in E, $P^\tau(x)$ is empty for all but finitely many τ in I. We sometimes abbreviate the fact that $i \epsilon P^\tau(x)$ by writing $i^\tau \epsilon x$, we say x <u>depends</u> on i^τ in such a case. If $i^\tau \not\epsilon x$, we say x is <u>independent</u> of i^τ. We can extend these definitions to a finite set S of elements of E, e.g., we let $P^\tau(S) = \bigcup_{x \epsilon S} P^\tau(x)$, mean by $i^\tau \epsilon S$ that $i \epsilon P^\tau(S)$ etc.

This concludes our basic definition of a many-sorted logical framework. We use the symbol "α" to mean any conjunctive element x such that $\phi(x)$ is finite; "β" for a disjunctive x such that $\phi(x)$ is finite; "γ" for any conjunctive x such that $\phi(x)$ is infinite; "δ" for any disjunctive x such that $\phi(x)$ is infinite. We write α_i for the i-th component of α, and similarly with β. We write $\gamma(i)$ for the i-th component of γ, and similarly with δ. Our conjugation laws imply that $\bar{\alpha}$ is some β, and that for any $i \le$ the number of terms of $\phi(\alpha)$, $\overline{\alpha_i} = (\bar{\alpha})_i$. Likewise for any γ, $\bar{\gamma}$ is some δ, and $\bar{\gamma}(i) = \overline{\gamma(i)}$.

Applications. In interpreting first order logic, the usual higher finite-ordered logics, and type theory, our set I will consist of types satisfying the following conditions:

1) 0 is a type.

2) Every type τ that is not 0 is of the form $(\tau_1,\ldots,\tau_n)$, where $\tau_1,\ldots,\tau_n$ are types.

In first order logic I contains 0 only. In second order logic I contains the types 0 and $(\underbrace{0,\ldots,0}_{n})$ for any positive integer n. (In general n-th order logic contains all the types satisfying 1) and 2) and having at most n-1 embeddings of parentheses.) In type theory I is the maximal set satisfying 1) and 2).

In the framework we construct, E is to the set of signed formulas. Our α's, β's, γ's and δ's are as in the individual logics, e.g., the α's will be all signed formulas of one of the four forms $T\ X \wedge Y$, $F\ X \vee Y$, $F\ X \supset Y$, $F \sim X$ etc. Our function ϕ shall assign to each α a two-term sequence $\langle\alpha_1, \alpha_2\rangle$; to each β a two-term sequence $\langle\beta_1, \beta_2\rangle$; to each γ a denumerable sequence $\langle\gamma(1), \gamma(2),\ldots,\gamma(n),\ldots,\rangle$, and to each δ a denumerable sequence $\langle\delta(1), \delta(2),\ldots,\delta(n),\ldots\rangle$, where the components α_i, β_i are obvious, and if γ (or δ) is a formula whose quantifier variable is of type τ, then $\gamma(i)$ ($\delta(i)$) is $\gamma(a_i)$ ($\delta(a_i)$, respectively), where a_i is the i-th parameter of type τ.

In first order logic the set of axioms is empty. In the other theories we are interested in, it consists of comprehension axioms like those we have seen in the second order logic.

The conjugation function is clear. If x is a formula of one of the logics in question, then $P^{\tau}(x) = \{i_1,\ldots,i_n\}$ if the i_1-st, i_2-nd,...,i_n-th parameters of type τ are found in the formula x.

Again we point out that the use of frameworks not only helps us to see the properties operating in proofs, but also allows us to prove theorems for a number of theories simultaneously. For instance, the primitives (~, ∧, etc.) could be replaced by any other complete set, we could consider languages of many sorts at the lowest level, or we could consider languages with formulas of denumerable length. We will not develop the theory of our frameworks any further, but will content ourselves with illustrating their usefulness by giving a proof of the Henkin completeness theorem (cf. Henkin [1]) which will thus simultaneously be a proof for all the finite-order logics and type theory. The definitions we proceed to give will be restricted to those needed to prove that theorem. We only remark that with further definitions the theory given in §§1, 2, 3 can be developed and results similar to those in Smullyan [2] can be obtained.

We let S be a subset of E.

We say S is closed downwards if for every conjunctive element c, if c is in S, then all components of c are in S, and for any disjunctive element d of S, at least one component of d is in S.

We call S a <u>truth set</u> if S is closed downwards, and S contains all the axioms, and for every element x, exactly one of the pair x, $\bar{x}$ lies in S.

We call x <u>valid</u> (in the framework $\mathcal{L}$) if x belongs to all truth sets, and <u>satisfiable</u> if x belongs to at least one. We call S (simultaneously) satisfiable if S is a subset of at least one truth set. (We note that x is valid iff $\bar{x}$ is not satisfiable.)

By a <u>tableau</u> for a set S (of elements of E) we mean a tree constructed as follows: We start the tree by taking any element of S as the origin. Now suppose $\mathcal{T}$ is a tableau for S and θ is any branch. Then we may extend θ by any of the following operations, the result again being a tableau for S:

(A) If C is a conjunctive term of θ, we may extend θ to (θ, C_i), where C_i is any component of C.

(B) For any term β of θ, we may simultaneously extend θ to $(\theta,\beta_1),\ldots,(\theta,\beta_n)$, where $\beta_1,\ldots,\beta_n$ are the components of β.

(C) For any δ in θ, we may extend θ to $(\theta,\delta(i))$, where i is such that for every τ, i^{τ} is independent of all the members of S and θ.

(D) For any axiom x, we can extend θ to (θ,x).

(E) For any s in S, we can extend θ to (θ,s).

(F) For any x in E, we can simultaneously extend θ to (θ,x), $(\theta,\bar{x})$.

A branch of a tableau is called <u>closed</u> if it contains some elements and its conjugate; otherwise it is called <u>open</u>. A tableau is called closed if all its branches are closed.

Now we define a set S to be syntactically inconsistent if there is a closed tableau for S. We define an element x to be provable if the unit set $\{\bar{x}\}$ is syntactically inconsistent, and by a proof of x we mean a closed tableau for $\bar{x}$.

We henceforth let S be any denumerable set (of elements of E) such that there are only finitely many i such that some element of S depends on i^{τ} for some τ.

Now we define a set M to be E-complete if for every $\delta \in M$, there is some i for which $\delta(i) \in M$. The Henkin completeness theorem is based on the following:

Lemma. If M is both E-complete and a maximal set which is syntactically consistent, then M is a truth set.

Proof. Let M be a maximal syntactically consistent set which is also E-complete. We wish to show M is a truth set. We note that saying M is syntactically consistent is the same as saying that no finite subset of M has a closed tableau.

Now consider an arbitrary element x of E. We first wish to see that exactly one of $x, \bar{x}$ is in M. It is clear that both of $x, \bar{x}$ cannot be in M because then the subset $\{x, \bar{x}\}$ of M would have a closed tableau. But if neither of them is in M, then for some finite subset M_1 of M, $M_1 \cup \{x\}$ has a closed tableau, and for another finite subset M_2 of M, $M_2 \cup \{\bar{x}\}$ has a closed tableau. But then, because of the cut rule, rule F, in the proof system, $M_1 \cup M_2$ has a closed tableau, which is impossible.

Now let a be an arbitrary axiom of E. Either a or $\bar{a}$ must be in M be the above. But $\bar{a}$ cannot be in M since $\{\bar{a}\}$ has a closed tableau because of the axiom rule, rule D.

It remains to see that M is closed downwards. Let C be a conjunctive element of M and let C_i be an arbitrary component of C. If C_i is not in M, then $\overline{C_i} = (\bar{C})_i$ is in M. But because of rule A, the set $\{C, [\bar{C}]_i\}$ has a closed tableau, so that $\overline{C_i}$ cannot be in M. For a disjunctive element D, the β case is similar to the case of a conjunctive element. If D is a δ, however, we must use the E-completeness of M, which gives us the desired fact immediately.

Now we wish to prove the main theorem of this paragraph.

Theorem 31 (proof based on one due to Henkin). (a) If a finite set S is syntactically consistent, it is satisfiable. (b) If an element x of E is valid, it is provable.

Proof. We will prove a, from which b follows easily. By the lemma we need only construct an E-complete maximal syntactically consistent set M containing S. We do this as follows:

Let $e_1,\ldots,e_n,\ldots$ be an enumeration of all the elements of E.

We let M_0 be the empty set. Now assume sets M_i have been defined for all $i < n$. At step n we let M_n be the following set:

(i) If $M_{n-1} \cup S \cup \{e_n\}$ is not syntactically consistent, we let $M_n = M_{n-1}$; otherwise:

(ii) If e_n is a δ, we let j be an integer such that e_n and all the elements of $M_{n-1} \cup S$ are independent of j^{τ} for every τ. Then we let $M_n = M_{n-1} \cup \{e_n, e_n(j)\}$. If neither the hypotheses of (i) or (ii) hold:

(iii) We let $M_n = M_{n-1} \cup \{e_n\}$.

Finally we let $M = \bigcup_i M_i$. We must show M is an E-complete maximal syntactically consistent set that contains S. By (ii) M is clearly E-complete. It is syntactically consistent because every M_i, $i = 1,2,\ldots,$ is, and every finite subset of M must lie entirely in one of the M_i. We must see that no proper superset of M is syntactically consistent. Assume to the contrary that M' is such a superset of M and that $m \in M' - M$. Let m be the element e_n of E. Since M' is syntactically consistent, clearly its finite subset $M_{n-1} \cup \{m\}$ is syntactically consistent and thus $m \in M_n \subset M$, which is a contradiction.

This completes the proof of the theorem (which actually can be generalized in a number of directions with slight changes in the proof). It is clear that some of what we feel to be significant properties of our logics are not needed for this proof. For instance the concept of degree is not used (this is partly because we call a tableau closed even when only a non-atomic formula and its conjugate appear on a branch). Also the properties of the axioms in the system are unimportant. If one thinks for a moment about the concepts and requirements that would have to be introduced to develop the material of §§1, 2, 3, and sees that a number of new ones would be needed (e.g., that for every element x of E there is an axiom corresponding to x in some sense) in addition to ones similar to those introduced by Smullyan in [2]. On the other hand, it seems that the Henkin proof is essentially a first order proof. We might point out, however, that this is because we used the truth set concept for our semantics, rather than that of a model. (Also, treating second order logic allowed us to overlook the difference between extensional and non-extensional theories.) For one can look at the difference between the Henkin completeness proofs usually given for different order logics as being one of going from a truth set to a model, which must be done in different ways for different orders because of the varied definitions of a model. This does not seem very significant when one considers the difference between first and second order logic, but models for type theory are quite a bit more involved (they have many more levels, orders!) than those of first order logic.

Chapter III — OTHER HIGHER ORDER SYSTEMS DUE TO SCHÜTTE

In this chapter we will continue our discussion of Schütte's Beweistheorie begun in Chapter I. We will introduce a number of higher order systems, prove their consistency and then give some indication as to how one might interpret classical analysis in such systems. Our systems will be of two fundamentally different kinds: type-free systems, which are propositionally incomplete, and systems of ramified type theory. We will omit Schütte's discussion of similar systems with descriptive operators.

§1. The System Λ of Type-free Logic

We give an abstraction term formulation for a system Λ of type-free logic.

The System Λ. The primitive symbols of the system Λ are:

1) a nonempty set of primitive constants;

2) a set of function symbols;

3) a denumerable list of variables $x_1, x_2, \ldots,$;

4) The logical symbols $\neg, \wedge, \vee, \supset, \forall, \exists$;

5) the operator λ;

6) The comma and parentheses.

As the constants, function symbols, and variables are not subdivided into sets of different kinds (as in second order logic, for instance), we say our system is a type-free system. The function symbols will be used in the development of mathematical theories within Λ.

We can now define the _terms_ and _formulas_ of Λ simultaneously as follows:

1) Every primitive constant, and variable is a term.

2) If ρ is a function symbol and $t_1,\ldots,t_n$ are terms

$$\rho(t_1,\ldots,t_n)$$

is also a term.

3) If A is a formula,

$$\lambda x_1\ldots x_n\ A$$

is a term $(n > 0)$.

4) If $t_1,\ldots,t_n,t$ are terms,

$$tt_1,\ldots,t_n$$

is a formula.

5) If A is a formula, so is $\neg A$, $(\exists x)A$ and $(\forall x)A$.

6) If A and B are formulas, so are $(A \vee B)$, $(A \wedge B)$, and $(A \supset B)$.

We sometimes write the formula $tt_1,\ldots,t_n$ (especially when t is a term by 3), a so-called _abstraction term_) as $t_1,\ldots,t_n \in t$. If the t in $tt_1,\ldots,t_n$ is a primitive constant, we say the formula is an _atomic formula_.

We define open and closed terms and signed formulas in the obvious way. Unless mention is made to the contrary, or the meaning intended is clear by the context, in the future we will consider only closed terms and formulas and only signed formulas. The α, β, γ, δ, and λ formulas are defined as in second order logic (although here we do not have to worry about the "kind" of the variable for the latter three).

We assume we are given a set Λ_f of constant atomic formulas such that for no formula X does both X and $\overline{X}$ belong to Λ_f.

In our tableau system for Λ, the rules for constructing a tableau for a set S are as follows:

<u>α rule</u>

$$\begin{array}{c} \alpha \\ | \\ \alpha_1 \end{array} \qquad \begin{array}{c} \alpha \\ | \\ \alpha_2 \end{array}$$

<u>β rule</u>

$$\begin{array}{ccc} & \beta & \\ / & & \backslash \\ \beta_1 & & \beta_2 \end{array}$$

<u>γ rule</u>

$$\begin{array}{c} \gamma \\ | \\ \gamma(t) \end{array} \quad t, \text{ any term}$$

<u>δ^ω rule</u>

$$\begin{array}{c} \delta \\ \wedge \cdots | \cdots \wedge \\ \delta(t) \end{array}$$

a branch for every term t

<u>λ rule</u>

$$\begin{array}{l} \pi\ c_1,\ldots,c_n \in \lambda x_1 \ldots x_n\ A(x_1,\ldots,x_n) \\ | \\ \pi\ A(c_1,\ldots,c_n) \end{array}$$

<u>Assumption Rule</u>

$$\begin{array}{l} S_1 \\ \vdots \\ S_n \end{array} \qquad S_i \in S, \; i = 1,\ldots,n$$

A branch in a tableau in Λ is closed if either

(a) it contains a formula X, where X is in Λ_f, or

(b) it contains some atomic formula X and its conjugate $\overline{X}$.

We note:

<u>Proposition 44</u>. There exists a constant atomic formula in Λ that does not have a closed tableau.

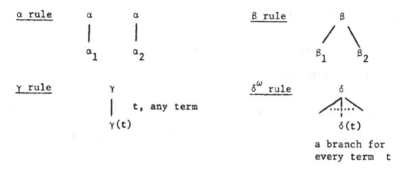

Proof. Since the set of primitive constants is nonempty, containing an element C, say, we can construct a constant atomic formula $T\ C \in C$. We claim either this formula or its conjugate do not have a closed tableau. For no rule of Λ is applicable to either formula. Clearly a tableau for one of them could not close by condition (b). But not both $T\ C \in C$ and $F\ C \in C$ can be in Λ_f, so one must not have a closed tableau.

The following proposition says something apparently more significant about the system.

Proposition 45. There is an (unsigned) formula in Λ that is propositionally true, but not provable, moreover, one of the form $A \vee \neg A$.

Proof. Let us denote the formula $(\lambda x(x \in x) \in \lambda x(x \in x))$ by U. It suffices to consider tableaux for $U \vee \neg U$ beginning

$$F\ U \vee \neg U$$

$$F\ U$$

$$F \neg U$$

$$T\ U$$

Only the λ rule is applicable to $F\ U$ and $T\ U$, however, and any such application just results in a formula already in the tableau. Thus we can never obtain a closed tableau for the formula in question.

We note that the proof of the above proposition shows us that it is not true that if a branch contains a formula X and its conjugate $\overline{X}$, it can be extended to a closed branch.

As in number theory (Ω) and second order logic (A_1^2) we can obtain the syntactic consistency of Λ by showing that the cut rule is a derivable rule of the system. Thus we now prove a quite restricted cut elimination theorem (this proof is an extention of the ones for first order logic and type theory).

Theorem 32. Let $\mathcal{T}$ be a closed tableau

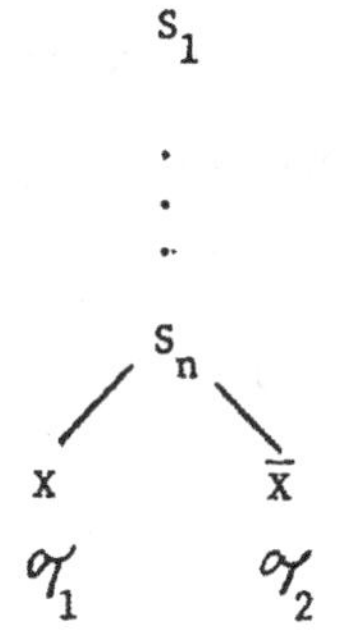

for a set $S = \{S_1,\ldots,S_n\}$, where the only cut in $\mathcal{T}$ occurs just after the assumption rule. We can transform $\mathcal{T}$ to a cut-free tableau.

Proof. We will prove this theorem by a triple transfinite induction.

We define the rank of a formula in a proof and of a proof as in Ω (see the introduction). If the rank of X is $\leq \alpha_1$ and the rank of $\overline{X}$ is $\leq \alpha_2$, we define the rank of the cut to be $\leq \alpha_1 \# \alpha_2$. We define the degree of an unsigned formula as follows:

a. Any formula of the form $tt_1,\ldots,t_n$, where $t,t_1,\ldots,t_n$ are terms is of degree 0.

b. If A is of degree n, $\neg A$, $(\forall x)A$, and $(\exists x)A$ are of degree $n+1$.

c. If A is of degree n and B is of degree m, $(A \wedge B)$, $(A \vee B)$ and $(A \supset B)$ are all of degree $n+m$.

The degree of a signed formula is the degree of its body and the <u>degree of a cut</u> is the degree of its cut formula.

We now define the <u>order</u> of a formula X in a tableau $\mathcal{T}$:

i. If X has no descendents in $\mathcal{T}$, then for every ordinal α, the order of X is $\leq \alpha$.

ii. If X has descendents in $\mathcal{T}$, let $\{X_i\}$ be the set of <u>direct</u> descendents of X in $\mathcal{T}$ and let the order of X_i be $\leq \alpha_i$. Then if X is <u>not</u> a λ formula, and μ is any ordinal such that for all i, $\alpha_i \leq \mu$, then the order of X is $\leq \mu$.
If X <u>is</u> a λ formula and μ is any ordinal such that for all i, $\alpha_i \leq \mu$, then the order of X is $\leq \mu+1$.

Thus the <u>order</u> of a formula measures how many successive λ rule applications it leads to. If a tableau $\mathcal{T}$ has a cut

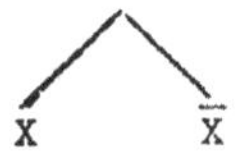

the order of X in $\mathcal{T}$ being $\leq \alpha_1$, the order of $\overline{X}$ in $\mathcal{T}$ being $\leq \alpha_2$, we say the <u>order of the cut</u> is $\leq \max\{\alpha_1, \alpha_2\}$.

Now assume a closed tableau $\mathcal{T}$, set S, cut to X, $\overline{X}$ satisfying the assumptions of the theorem.

We take as induction hypotheses that the theorem holds whenever the cut in question has either

a. order $\le \rho$, degree $\le g$, and rank $< \alpha_1 \# \alpha_2$, or

b. order $\le \rho$ and degree $< g$ (rank arbitrary), or

c. order $< \rho$ (degree and rank arbitrary).

It is not too difficult to see that it suffices to prove the theorem for a cut of order $\le \rho$, degree g, and rank $\le \alpha_1 \# \alpha_2$ under these hypotheses.

So let $\mathcal{T}$ be the closed tableau

$$\begin{array}{ccc} & S_1 & \\ & \vdots & \\ & S_n & \\ / & & \backslash \\ X & & \overline{X} \\ \mathcal{T}_1 & & \mathcal{T}_2 \end{array}$$

where $\mathcal{T}_1$ and $\mathcal{T}_2$ are cut-free, and the cut to X and $\overline{X}$ is of order $\le \rho$, degree g, and rank $\le \alpha_1 \# \alpha_2$. We must consider several cases:

(1) Let us assume either $\mathcal{T}_1$ or $\mathcal{T}_2$ is empty, say $\mathcal{T}_1$. Then if X is not needed to close its branch,

$$(*) \qquad \begin{array}{c} S_1 \\ \vdots \\ S_n \end{array}$$

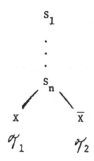

is itself a closed cut-free tableau for S. If X is needed to close its branch, we must consider the closure condition used. If the branch closes by condition a, i.e., X is a constant atomic formula in Λ_f, then $\overline{X}$ is a constant atomic formula that cannot be in Λ_f. Thus if $\overline{X}$ is needed to close its branch (if not

$$(**)\qquad \begin{array}{c} S_1 \\ \cdot \\ \cdot \\ \cdot \\ S_n \\ \mathcal{T}_2 \end{array}$$

is a closed cut-free tableau for S), it must be by condition b, in which case $X \in S$ and (*) is again a suitable tableau. Finally if X is needed to close its branch by condition b, $\overline{X} \in S$ and (**) is a suitable tableau.

(2) Let us assume that neither $\mathcal{T}_1$ nor $\mathcal{T}_2$ is empty, but that for at least one of them, say $\mathcal{T}_1$, the first rule applied is not applied to the cut formula above the tree. Thus we can write our tableau

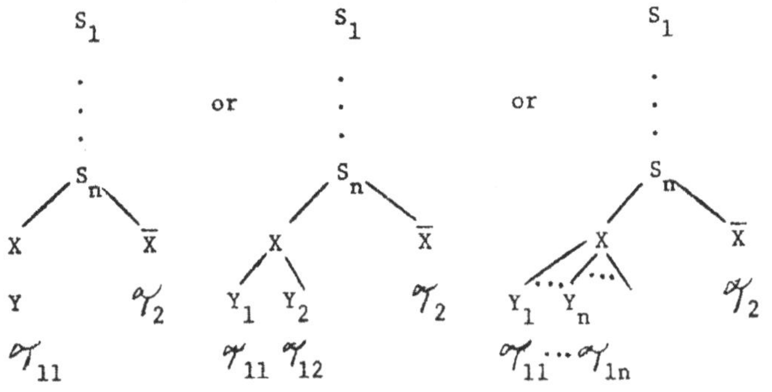

We transform such a tableau to a tableau

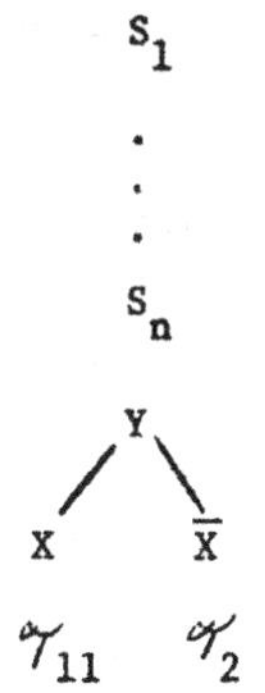

or

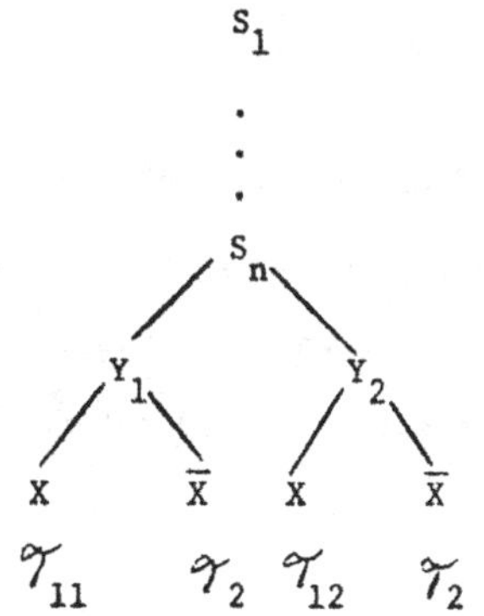

or

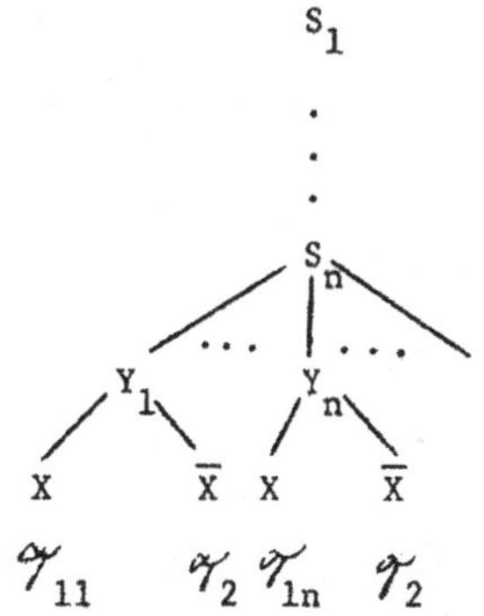

respectively. Let us consider the second situation, the two others being similar.

Now we consider the tableaus

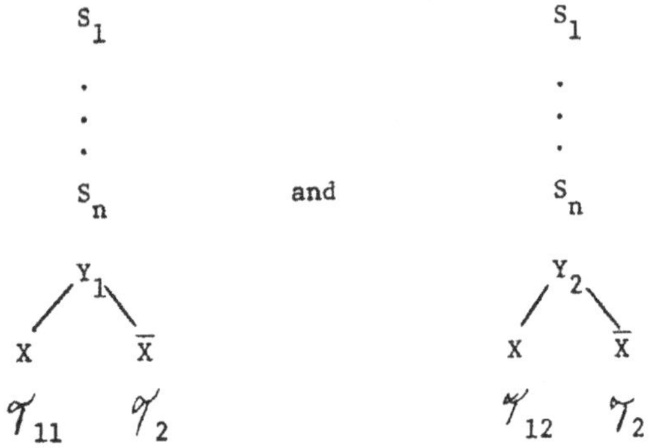

as tableaus for $S^1 = S \cup \{Y_1\}$ and $S^2 = S \cup \{Y_2\}$. Then their cuts have order $\leq \rho$, degree $\leq g$, but a rank $< \alpha_1 \# \alpha_2$, (where the rank of our original cut was $\leq \alpha_1 \# \alpha_2$). We can thus apply the induction hypothesis to obtain closed cut-free tableaus

$$\begin{array}{c} S_1 \\ \vdots \\ S_n \\ Y_1 \\ \mathcal{T}_1' \end{array} \qquad \text{and} \qquad \begin{array}{c} S_1 \\ \vdots \\ S_n \\ Y_2 \\ \mathcal{T}_2' \end{array}$$

It is easy to see then that

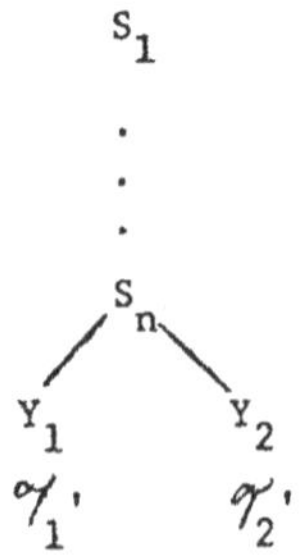

is a closed cut-free tableau for S.

(3) Let us assume that neither $\mathcal{T}_1$ nor $\mathcal{T}_2$ is empty and that the first rule applied in both $\mathcal{T}_1$ and $\mathcal{T}_2$ is applied to the cut formula above the tree. We consider the different forms of the cut formula separately.

(3a) Let X be an α, $\overline{X}$ thus a β, and assume α_1 is the formula first obtained from α. So we can write our tableau

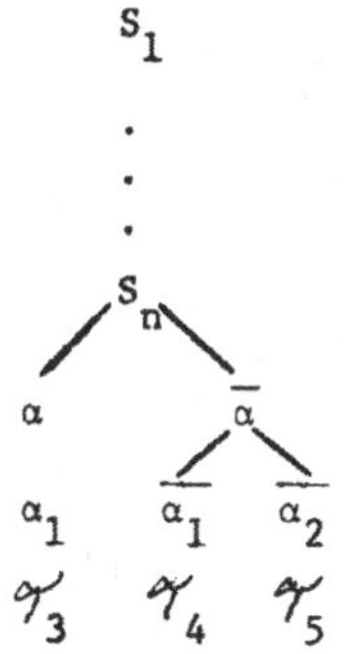

We transform this tableau to the tableau

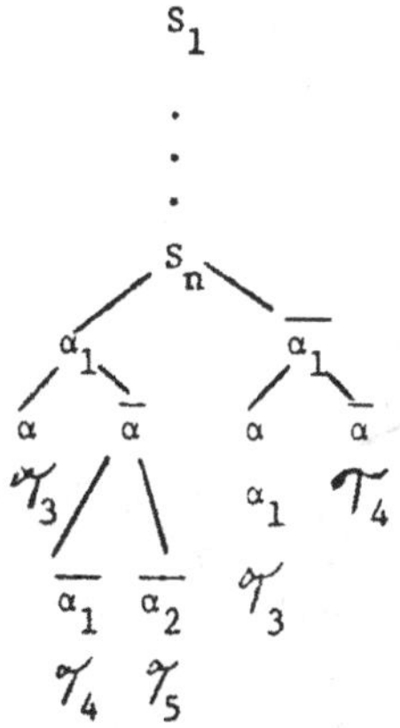

We see that the upper-most cut is of lower degree, but the same order as the original one, while the two lower cuts are of the same degree and order, but of a lower rank than the original. We can proceed as in the last case to use our induction hypotheses to eliminate these cuts and obtain a tableau satisfying our requirements.

(3b) Let X be a γ, $\overline{X}$ thus a δ. We can write our tableau

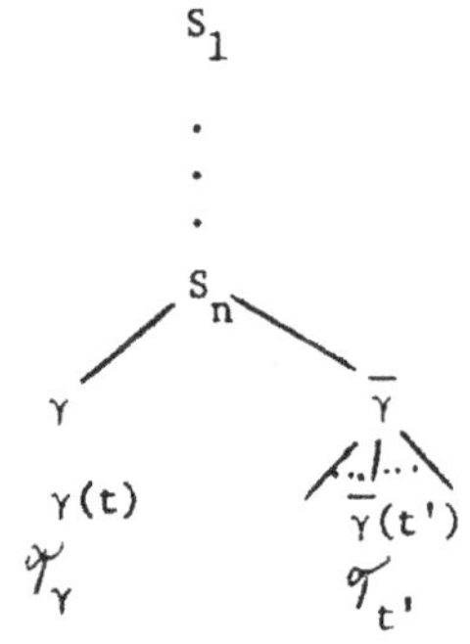

We can transform this tableau to

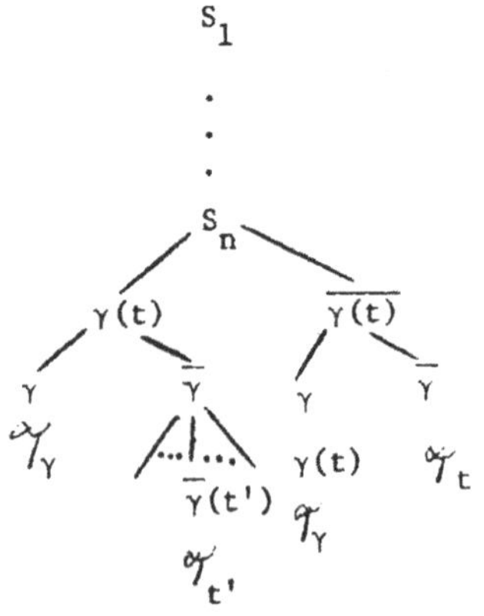

The rest of this case is like that of (3a).

(3c) Let X and $\overline{X}$ be λ formulas. Thus we can write our tableau

$$S_1$$

$$\vdots$$

$$S_n$$

$$\pi c_1,\ldots,c_n \in \lambda x_1 \ldots x_n A(x_1,\ldots,x_n) \qquad \overline{\pi} c_1,\ldots,c_n \in \lambda x_1 \ldots x_n A(x_1,\ldots,x_n)$$

$$\pi\ A(c_1,\ldots,c_n) \qquad \overline{\pi}\ A(c_1,\ldots,c_n)$$

$$\mathcal{T}_1 \qquad \mathcal{T}_2$$

For convenience let us assume that the λ rule is never again applied to X or $\overline{X}$ in $\mathcal{T}_1$ or $\mathcal{T}_2$ (for it would certainly be unnecessary, and, in any case, we could eliminate all such repetitious uses in a constructive manner). Then we can transform the given tableau to

$$S_1$$

$$\vdots$$

$$S_n$$

$$\pi A(c_1,\ldots,c_n) \qquad \overline{\pi} A(c_1,\ldots,c_n)$$

$$\mathcal{T}_1 \qquad \mathcal{T}_2$$

Here the cut involved may have higher degree, but does have a lower order than the original one. Thus we can again apply our induction hypotheses to obtain a cut-free closed tableau for S.

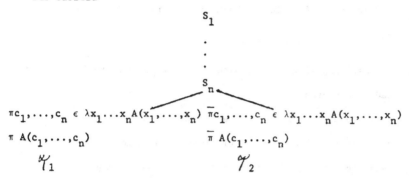

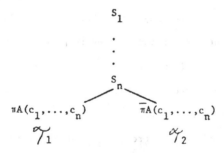

As usual we have as a corollary of the cut elimination theorem the <u>syntactic consistency</u> of our system:

<u>Theorem 33</u>. There is no formula A such that both A and $\neg A$ are provable in Λ.

In this case proposition 45 provides us with the unprovable formula needed in the proof.

With a little more analysis, we can actually obtain further constructive information from our cut elimination theorem, namely another "consistency" theorem. As the results we will discuss hold for all the systems we deal with, we will present the material in terms of an arbitrary system satisfying the relevant requirements.

Thus let Δ be a tableau system such that:

Δ1) the rules of Δ include the α and β rules;

Δ2) the cut rule is eliminable in Δ.

We note that such a system may not be "propositionally complete" because we are not stipulating that every pair of conjugate formulas lead to a closed branch. We define for any signed formula X in Δ the propositional decomposition tree for X as follows:

a. X is the origin of the tree.

b. If α is a point in the tree, α_1 and α_2 are left and right successors of α in the tree.

c. If β is a point in the tree, β_1 and β_2 are the left and right successors of β in the tree.

We say the bodies of the end points of the propositional decomposition tree for X are the <u>Boolean atoms</u> of X. Clearly each such formula corresponds to a particular subformula in X. We say a (signed or unsigned) formula of Δ is a <u>tautology</u>, or is <u>propositionally true</u>, when it is true under every assignment of truth values to its Boolean atoms (according to the usual truth definition associated with the propositional connectives). Similarly, we say a formula of Δ is <u>propositionally false</u> when it is false under every such assignment.

We define a <u>complete propositional tableau</u> for a formula X in Δ as might be expected. It is a finite tableau $\mathcal{T}_X$ in which only α and β rules are used and in which, for each branch θ in it, for every α in θ, both α_1 and α_2 are in θ, and for every β in θ, either β_1 or β_2 occurs in θ. We recall that the propositional completeness theorem tells us: <u>If X is a propositionally false signed formula in Δ and θ is a branch in a complete propositional tableau for X, θ contains some formula πB and its conjugate, where B is a Boolean atom of X.</u>

Now let X be an arbitrary tautology in Δ and let $\mathcal{T}_X$ be a complete propositional tableau for X. Let $\theta_1,\dots,\theta_i$ be the branches in $\mathcal{T}_X$, and let S_j be the set of formulas in θ_j that are not α's or β's ($1\le j\le i$). It is easy to see that the body of any formula in an S_j is a Booleon atom of X. It can be shown by induction that <u>for any truth assignment to the Booleon atoms of X, X is true iff either all the formulas in S_1 are true, or all the formulas in S_2 are true or ..., or all the formulas in S_i are true</u>. Let the elements of S_j be denoted by $\pi_1^j s_1^j,\dots,\pi_{n_j}^j s_{n_j}^j$. Then let t_k^j be s_k^j if π_k^j is T and let t_k^j be $\neg s_k^j$ if

π_k^j is F. Finally let S_j' be $t_1^j, \ldots, t_{nj}^j$ $(1 \le j \le i)$. Let Y_k^j denote Ts_k^j if t_k^j is $\neg s_k^j$ (i.e., Fs_k^j is in S_j) and denote Fs_k^j otherwise (i.e., $T\ s_k^j$ is in S_j). We can now consider a particular complete propositional tableau for $F\ \hat{S}_1' \vee \ldots \vee \hat{S}_i'$ [1] (for convenience we assume repetitions and α and β rules allowing arbitrarily many components in conjunctions and disjunctions).

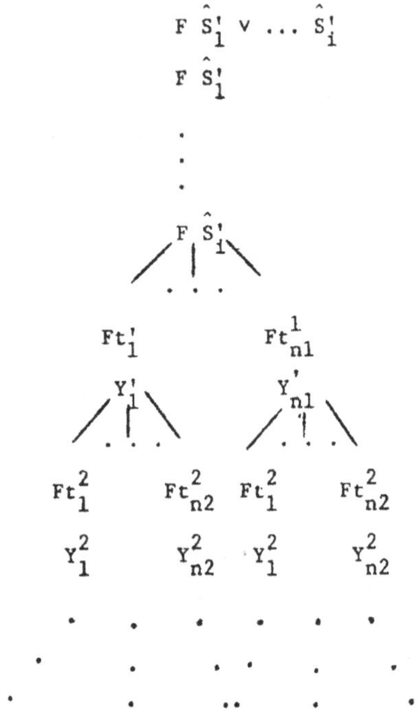

We see that each branching is to the conjugates of all the elements in some branch in $\mathcal{T}_X$ (i.e., $Y_1', \ldots, Y_{n1}'$ are the conjugates of the formulas in θ_1 etc.). Thus for each way of choosing one formula from each θ_j, our tableau for $F\ \hat{S}_1' \vee \ldots \vee \hat{S}_i'$ has a branch whose non-α and non-β formulas are precisely the conjugates of the chosen formulas. But since $\hat{S}_1' \vee \ldots \vee \hat{S}_i'$ is a tautology, each such set of formulas con-

[1] If S is a set of unsigned formulas, $\hat{S}$ is the conjunction of the elements of S.

a formula and its conjugate. Thus we can conclude about $\mathcal{T}_X$: <u>For each set C formed by choosing one formula from each</u> S_j, <u>C contains some formula and its conjugate</u> (whose bodies are Booleon atoms of X).

Let us now state the theorem we wish to prove (the property of a system stated in the theorem is closely related to Schütte's "regularity").

<u>Theorem 34</u>. Let S be a finite set of formulas of Δ and let X be a tautology not in S. If $S \cup \{X\}$ has a closed tableau $\mathcal{T}$ in a system Δ satisfying conditions $\Delta 1$ and $\Delta 2$, we can transform $\mathcal{T}$ into a closed tableau for S.

Proof. Let us be given a set S, formula X, and tableau $\mathcal{T}$ satisfying the conditions of the theorem. We first transform our tableau $\mathcal{T}$ into a tableau of the form

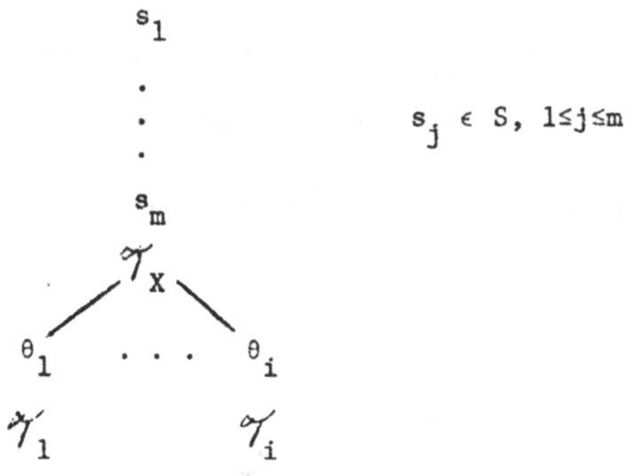

where $\mathcal{T}_X$ is a complete propositional tableau for X with branches $\theta_1, \ldots, \theta_i$. We do this as follows. Let us assume at some step in the process we have a tableau of the form

$$
(*) \qquad
\begin{array}{ccc}
 & s_1 & \\
 & \vdots & \\
 & s_m & \\
 & \mathcal{T}^0_X & \\
 / & \cdots & \backslash \\
\theta_1 & & \theta_i \\
\mathcal{T}_1 & & \mathcal{T}_i
\end{array}
$$

where $\mathcal{T}^0_X$ is a tree with origin X in which only α and β rules are used and all formulas are descendents of X, while if an α or β rule is applied to a formula Y in $\mathcal{T}^0_X$ to obtain a formula in $\mathcal{T}^0_X$, the rule is never applied to Y to obtain a formula in any of the $\mathcal{T}_1, \ldots, \mathcal{T}_i$. If every α or β formula in $\mathcal{T}^0_X$ has the relevant rule applied to it on every branch through it in $\mathcal{T}^0_X$, we stop, for $\mathcal{T}^0_X$ must be a complete tableau for X. Otherwise let Y be an α or β formula in $\mathcal{T}^0_X$ and let $\xi_1, \ldots, \xi_n$ be the branches in $\mathcal{T}^0_X$ on which the relevant rule has not been applied - we let $\mathcal{T}_{\xi_1}, \ldots, \mathcal{T}_{\xi_n}$ be the trees beneath them in (*). We change the part of the tree

$$
(*j) \qquad
\begin{array}{c}
s_1 \\
\vdots \\
s_m \\
\xi_j \\
\mathcal{T}_{\xi_j}
\end{array}
$$

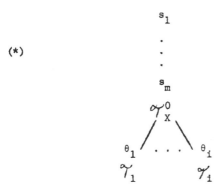

as follows: If Y is an α, we let $\mathcal{T}^-_{\xi_j}$ be $\mathcal{T}_{\xi_j}$ with all the Y_1's and Y_2's that occur in it deleted. Then we replace (*j) by

$$\begin{array}{c} s_1 \\ \cdot \\ \cdot \\ \cdot \\ s_m \\ \xi_j \\ Y_1 \\ Y_2 \\ \mathcal{T}^-_{\xi_j} \end{array}$$

By this we mean we replace $\mathcal{T}_{\xi_j}$ in our given tableau by Y_1, Y_2, $\mathcal{T}^-_{\xi_j}$) If Y is a β, we let $\mathcal{T}^1_{\xi_j}$ ($\mathcal{T}^2_{\xi_j}$) be the result of first eliminating all the subtrees under a Y_2 (Y_1, respectively) in it, and then eliminating all the Y_1's and Y_2's. We then replace (*j) by

$$\begin{array}{ccc} & s_1 & \\ & \cdot & \\ & \cdot & \\ & \cdot & \\ & s_m & \\ & \xi_j & \\ \swarrow & & \searrow \\ Y_1 & & Y_2 \\ \mathcal{T}^1_{\xi_j} & & \mathcal{T}^2_{\xi_j} \end{array}$$

It is easy to see that the final result of this process is a tableau

(*)

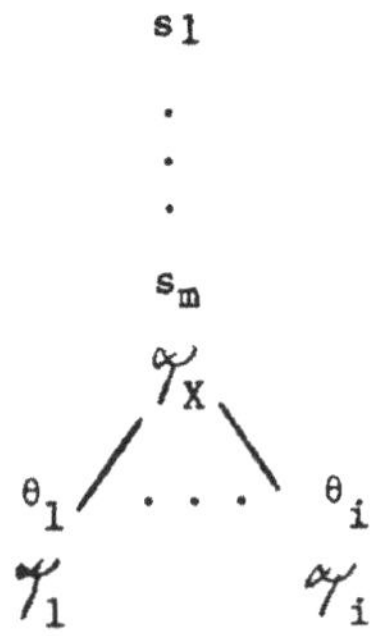

where $\mathcal{T}_X$ is a complete propositional tableau for X with branches $\theta_1,\ldots,\theta_i$, and where every descendent of X in the $\mathcal{T}_1,\ldots, \mathcal{T}_i$ is a descendent of some πB, where B is a Booleon atom in X.

Now let $B_1,\ldots,B_n$ be the Booleon atoms of X. Let $\pi_1 B_1,\ldots,\pi_n B_n$ be the result an arbitrary assignment of signs to these formulas. Let S_j be the set of atomic formulas (all of which are signed Boolean atoms of X) in the sub-branch θ_j of (*), $j = 1,\ldots,i$. We claim there is some j such that S_j is a subset of $\{\pi_1 B_1,\ldots,\pi_n B_n\}$. This is a consequence of a fact established earlier, namely, that any set C formed by choosing one formula from each S_j must contain some formula πB and its conjugate, where B is a Boolean atom of X. Perhaps the easiest way to see this is as follows. Clearly a given S_j might contain neither $T\ B_k$ nor $F\ B_k$ for some k. Let us replace such a set S_j by all the sets formed as follows: Let $A_1,\ldots,A_\ell$ be the Boolean atoms of X such that neither $T\ A_k$ nor $F\ A_k$ occurs in S_j, $k = 1,\ldots,\ell$, and let $\pi_1,\ldots,\pi_\ell$ be an arbitrary list of ℓ signs. Then $S_j \cup \{\pi_1 A_1,\ldots,\pi_\ell A_\ell\}$ is one of the sets in question. Let us denote the sets replacing S_j

by $S_j^1,\dots,S_j^{nj}$. It is not difficult to see that the sets S_j^k still possess the property that any set C formed by choosing one formula from each set must contain some formula πB and its conjugate, where B is a Boolean atom of X. And clearly if we can show that all the elements of a certain S_j^k are among the $\pi_1B_1,\dots,\pi_nB_n$ (i.e., that S_j^k is the same as $\{\pi_1B_1,\dots,\pi_nB_n\}$), we will know that all the elements of S_j are among the $\pi_1B_1,\dots,\pi_nB_n$, which is the assertion we are now trying to demonstrate.

Clearly some S_j^k must contain π_1B_1, for if every S_j^k contained $\overline{\pi}_1B_1$, we could choose that formula from each set to form a set C that did not contain some formula and its conjugate. Now assume we know that there is at least one set that contains $\pi_1B_1,\dots,\pi_\ell B_\ell$, $1\le\ell<n$. Among the sets that contain those formulas there must be at least one that also contains $\pi_{\ell+1}B_{\ell+1}$. For if not we could form a set C as follows. From the sets containing $\pi_1B_1,\dots,\pi_\ell B_\ell$, we choose $\overline{\pi}_{\ell+1}B_{\ell+1}$. From all other sets we choose either $\overline{\pi}_1B_1$ or $\overline{\pi}_2B_2$, or ..., or $\overline{\pi}_\ell B_\ell$. Such a set clearly contains no formula and its conjugate. This proves our assertion, which was, as we recall, that for any set $\{\pi_1B_1,\dots,\pi_nB_n\}$, where $B_1,\dots,B_n$ are the Boolean atoms in X, some S_j is the subset of that set, where S_j was the set of Boolean atoms from X in the branch θ_j in our tableau

(**)

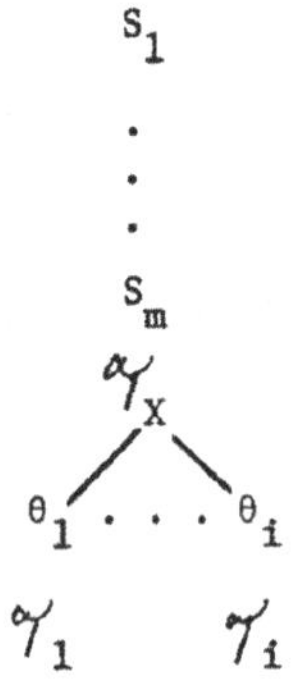

for $S \cup \{X\}$ - here $\mathcal{T}_X$ was a complete propositional tableau for X and all the descendents of X in the $\mathcal{T}_1, \ldots, \mathcal{T}_i$ were the descendents of formulas πB in $\mathcal{T}_X$, where B was a Boolean atom in X.

Using cuts, we can now replace the tableau (**) by a tableau

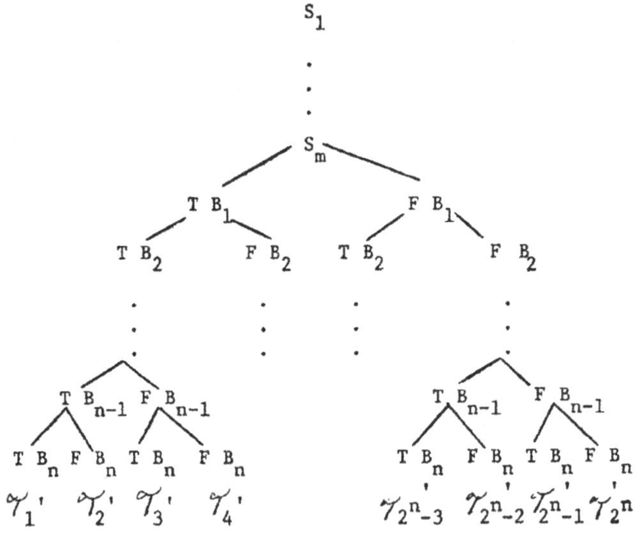

where the $\mathcal{T}_j'$ are chosen as follows: if the j-th branch ending in some πB_n is

$$
\begin{array}{c}
S_1 \\
\vdots \\
S_m \\
\pi_1 B_1 \\
\vdots \\
\pi_n B_n
\end{array}
$$

we let $\mathcal{T}'_j$ be any tree $\mathcal{T}_\ell$ from among the $\mathcal{T}_1, \ldots, \mathcal{T}_i$ such that S_ℓ is a subset of $\{\pi_1 B_1, \ldots, \pi_n B_n\}$.

We thus have a tableau for S with finitely many cuts. By the cut elimination theorem we can transform this tableau into a cut-free tableau for S. This completes the proof of the theorem.

We say that a system is <u>propositionally consistent</u> if no signed tautology has a closed tableau (which implies that <u>no propositionally false unsigned formula can be proved</u>, for if A is propositionally false, F A is a tautology).

If the system A possesses the further property:

Δ3) there is a formula that is not provable in Δ , we have finally:

<u>Theorem 35</u>. Any system Δ satisfying conditions Δ1, Δ2, and Δ3 is propositionally consistent.

Proof. By condition Δ3 there is some formula in Δ that is not provable. Thus there is some formula X that does not have a closed tableau. But if there were a tautology Y with a closed tableau

Y

then

X

Y

would be a closed tableau for $\{X\} \cup \{Y\}$, which, by theorem 34, could be transformed into a closed tableau for $\{X\}$, contrary to assumption.

Since clearly Λ satisfies conditions $\Delta 1$, $\Delta 2$, and $\Delta 3$, we have as a corollary:

<u>Theorem 36 (propositional consistency of Λ)</u>. No tautology has a closed tableau in Λ (and hence no propositionally false formula is provable in Λ).

We might mention why we did not carry out this discussion, which clearly delas only with "propositional" concepts, when we were considering Ω, for that system also clearly satisfies conditions $\Delta 1$, $\Delta 2$, and $\Delta 3$ and is thus propositionally consistent. The reason is that Ω also has the property that any branch containing some formula and its conjugate can be extended to a closed branch. This, along with $\Delta 1$ makes Ω propositionally complete, i.e., if X is a propositionally false signed formula, a complete tableau for X is closed. In such a case the results of theorems 34 and 35 can be seen almost immediately. For let Δ be a system that satisfies $\Delta 1$, $\Delta 2$, and $\Delta 3$, and is also propositionally complete. Let S be a finite set of formulas and let X be a tautology, and let $S \cup \{X\}$ have a closed tableau:

$$\begin{array}{c} S_1 \\ \cdot \\ \cdot \\ \cdot \\ S_n \\ X \\ \mathcal{T} \end{array}$$

Then, since $\overline{X}$ is propositionally false it has a closed tableau

$$\begin{array}{c} \overline{X} \\ \mathcal{T}' \end{array}$$

and S will have the tableau

$$\begin{array}{ccc} & S_1 & \\ & \cdot & \\ & \cdot & \\ & \cdot & \\ & S_n & \\ \swarrow & & \searrow \\ X & & \overline{X} \\ \mathcal{T} & & \mathcal{T}' \end{array}$$

whose single cut can be eliminated by $\Delta 2$. The assertion of theorem 35 then follows as before. Thus we did not need the more complicated discussion we gave for theorem 34 until we wanted to investigate the propositional consistency of Λ.

This completes our preliminary discussion of the system Λ. We will now define and discuss a different system of higher order logic, which is consistent and propositionally complete, but is unfortunately quite a bit more complex than Λ.

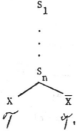

§2. The System T of Ramified Type Theory

We will now describe a system of type theory that can be proven constructively to be consistent and propositionally complete. To be able to accomplish this we will do two things we did not do in Λ. Firstly, we will make a type distinction (as in first and second order logic), so that for every predicate only arguments of certain types may belong to it. Also we will not allow "impredicative" concept building. This means that a concept that is defined by reference to a collection will never itself belong to that collection. For this purpose we will make a classification of the formal expressions so that every expression that contains quantifiers of a particular class will be in a higher class.

Thus before defining the system itself, we will first define our "types" and "classes". The classes will be ordinals and the types will be strings of symbols satisfying the following:

$\tau 1$. Each type belongs to some class.

$\tau 2$. The ordinal 0 is a type of the class 0.

$\tau 3$. If $\tau_1,\ldots,\tau_n$ are types $(n \geq 1)$ whose classes are ordinals smaller than an ordinal σ, then $\sigma(\tau_1,\ldots,\tau_n)$ is a type of the class σ.

The type 0 will be used to designate the base domain, while a type $\sigma(\tau_1,\ldots,\tau_n)$ will be the type of an n-place predicate of the class σ whose argument places belong to the types $\tau_1,\ldots,\tau_n$.

We define a non-total order relation for the types as follows. If $\sigma_1(\tau_1,\ldots,\tau_n)$ is a type and σ_2 is a larger ordinal than σ_1, then

$$\sigma_1(\tau_1,\ldots,\tau_n) < \sigma_2(\tau_1,\ldots,\tau_n).$$

Let us now proceed to define the system T of ramified type theory.

The primitive symbols of T are as in Λ (i.e., primitive constants, function symbols, variables, $\neg$, $\wedge,\vee,\supset,\forall,\exists,\lambda$, parentheses and comma), but every primitive constant and variable is of a particular type. We will assume denumerably many variables of each type. Every function symbol has associated with it an integer > 0 that is its "number of places." The function symbols can be used to develop a mathematical theory within T. Here we will use only functions whose arguments and values range over the base domain. For this reason we restrict the function symbols and the terms built with function symbols to the type 0.

The terms and formulas of the system T are defined as in the system Λ with the difference that here every term is assigned a particular type and every formula is assigned a particular class. We now give a simultaneous definition of the terms with types and of the formulas with their classes.

t1. Every primitive constant and every variable of type τ is a term of type τ.

t2. If ρ is an n-place function symbol and $t_1,\ldots,t_n$ are terms of type 0,

$$\rho(t_1,\ldots,t_n)$$

is also a term of type 0.

t3. If $A(x_1,\ldots,x_n)$ is a formula of the class σ_0 with variables $x_1, \ldots, x_n$ of the types $\tau_1, \ldots, \tau_n$, whose classes are $\sigma_1, \ldots, \sigma_n$, then

$$\lambda x_1 \ldots x_n A(x_1,\ldots,x_n)$$

is a term of the type

$$\sigma(\tau_1,\ldots,\tau_n),$$

where σ is the maximum of $\sigma_0, \sigma_1 + 1, \ldots, \sigma_n + 1$.

F1. If $t_1, \ldots, t_n$ and t are terms of the types $\mathcal{S}_1, \ldots, \mathcal{S}_n$ and $\sigma(\tau_1,\ldots,\tau_n)$, with $\mathcal{S}_1 \leq \tau_1, \ldots, \mathcal{S}_n \leq \tau_n$, then $tt_1,\ldots,t_n$ (often written $t_1,\ldots,t_n \in t$) is a formula of the class σ.

F2. If A is a formula of the class σ, $\neg A$ is also a formula of the class σ.

F3. If A and B are formulas, $(A \wedge B)$, $(A \vee B)$, and $(A \supset B)$ are also formulas. Their class is the maximum of the classes of A and B.

F4. If $A(x)$ is a formula of the class σ_0 with a variable x of type τ_1 and class σ_1, then

$$(\forall x)A(x) \quad \text{and} \quad (\exists x)A(x)$$

are formulas. Their class is the maximum of σ_0 and $\sigma_1 + 1$.

(In t2, t3, and F1, n must be ≥ 1.)

We note that the requirement on a type $\sigma(\tau_1,\ldots,\tau_n)$ that the classes $\sigma_1, \ldots, \sigma_n$ of the types $\tau_1, \ldots, \tau_n$ be smaller than σ is maintained in the above definition. For in t3 σ is $\geq \max\{\sigma_1 + 1,\ldots,\sigma_n + 1\}$. Also, the second requirement set forth at the beginning of this paragraph is taken into account when the class σ of formulas defined under F4 is larger than the class σ_1 of the quantified variable x.

We next note a simple corollary of the definition of the terms and formulas.

Proposition 46. If a term of type τ_0 contained in a term t of type τ (in a formula F of class σ) is replaced by another term of a type $\leq \tau_0$, the term t goes into a term of type $\leq \tau$ (the formula F goes into a formula of a class $\leq \sigma$).

The proof is by complete induction on the length of the string t (F). For a term t defined by t1 the assertion is trivial. In all other definition rules terms (formulas) of shorter lengths are assumed. For these the assertion holds by the induction hypothesis. By examining the particular definition requirements one can see that the assertion then also holds for the term (formula) being defined.

Strengthening of proposition 46. If the replaced term belongs to a smaller class than t (F), the type of t (the class of F) is not changed under the replacement.

The proof corresponds to that of proposition 46. For a term defined by requirement t1 the assertion is trivial. For the terms (formulas) that are assumed in other definition clauses, we have: a) If an assumed expression belongs to a smaller class than the defined expression, it is meaningless as far as the class specification of the defined expression. b) If the assumed expression belongs to the same class as the defined expression, the assertion holds for it by the induction hypothesis. The assertion then also carries over to the defined expression.

We assume the usual definitions for closed terms and formulas, for signed formulas, etc. From now on we will speak only of closed terms and formulas unless the context makes it clear (in some inductive definition) that we also wish to consider open formulas.

We say a formula of the form

$$t_1, \ldots, t_n \in t$$

is an <u>atomic</u> formula, if t is a primitive constant.

Let us now define the <u>tableau system for T</u> (it is essentially like that of Λ expect that certain type restrictions are made). Let us call a γ or δ formula with an outermost quantified variable of type τ a γ^τ or δ^τ formula, and let us denote a term of type τ by t^τ. Again we assume we are given a set of signed atomic formulas, which we call T_f. Now the rules for constructing a tableau in T for a set S are as follows:

<u>α rule</u>

$$\begin{array}{c} \alpha \\ | \\ \alpha_1 \end{array} \qquad \begin{array}{c} \alpha \\ | \\ \alpha_2 \end{array}$$

<u>β rule</u>

$$\begin{array}{ccc} & \beta & \\ \swarrow & & \searrow \\ \beta_1 & & \beta_2 \end{array}$$

<u>γ^τ rule</u>

$$\begin{array}{c} \gamma^\tau \\ | \\ \gamma^\tau(t^{\tau'}) \end{array}, \quad \tau' \leq \tau$$

<u>δ^τ rule</u>

$$\begin{array}{c} \delta^\tau \\ \swarrow \ | \ \searrow \\ \delta^\tau(t^{\tau'}) \end{array}$$

for all terms $t^{\tau'}$ such that $\tau' \leq \tau$

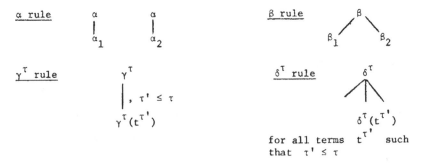

λ rule

$$\pi c_1,\ldots,c_n \in \lambda x_1\ldots x_n A(x_1,\ldots,x_n)$$
$$|$$
$$\pi\ A(c_1,\ldots,c_n)$$

Assumption Rule

$$\begin{array}{l} s_1 \\ \vdots \quad s_i \in S, \\ \quad\ i = 1,\ldots,n \\ s_n \end{array}$$

A branch in a tableau in T is closed if either:

(a) it contains a formula X from T_f, or

(b) it contains some atomic formula X and its conjugate $\overline{X}$.

In order to be able to prove properties about formulas inductively, we associate with every formula of the system T a natural number called its "coefficient." The coefficient of a formula F of the class σ will express how many formulas of the class σ are needed to build up the formula F according to the definition rules for formulas.

Definition of the coefficient of a formula:

b0. Every atomic formula has the coefficient 0.

b1. Let F be a formula

$$(t_1,\ \ldots,\ t_n \in \lambda x_1\ldots x_n A(x_1,\ldots,x_n))$$

by F1 that is of the class σ and contains a term

$$\lambda x_1\ldots x_n A(x_1,\ldots,x_n)$$

by t3. Let the formula

$$A(x_1,\ldots,x_n)$$

assumed by t3 have the class σ_0 and the coefficient b. Then we specify:

1. If $\sigma_0 < \sigma$, F has the coefficient 0.

2. If $\sigma_0 = \sigma$, F has the coefficient b + 1.

b2. If F is a formula $\neg A$ by F2, where A has the coefficient b, then F has the coefficient b + 1.

b3. If F is a formula $(A \wedge B)$, $(A \vee B)$ or $(A \supset B)$ by F3, we first define a number b.

1. If A and B belong to different classes, let b be the coefficient of the formula A or B that belongs to the largest class.

2. If A and B belong to the same class, let b be the maximum of the coefficients of A and B.

In both cases F has the coefficient b + 1.

b4. Let F be a formula $(\forall x)A(x)$ or $(\exists x)A(x)$ of the class σ built by F4. Let the formula A(x) assumed in the definition F4 have the class σ_0 and the coefficient b. Then:

1. If $\sigma_0 < \sigma$, F has the coefficient 0.

2. If $\sigma_0 = \sigma$, F has the coefficient b + 1.

These cases include all formulas. The formulas defined by F1 fall under b0 or b1, and all other formulas fall under b2 - b4. The clauses b0 - b4 define the coefficient <u>constructively</u>, since their statements always depend only on the coefficients of shorter formulas.

<u>Proposition 47</u>. If a term of type τ_0 with class $< \sigma$ occurring in a formula F of class σ is replaced by another term of a type $\leq \tau_0$ the formula F goes into a formula with the same coefficient.

Proof. The proof is by complete induction on the length of the formula F. By the strengthening of proposition 46, F goes into a formula F* of the same class under the replacement.

I. If F is an atomic formula $(t_1,\ldots,t_n \in t)$, by F1 t belongs to the same class σ as the formula F. Thus no replacement takes place in t. Then F* is also an atomic formula and hence has the coefficient 0 just as F did.

II. In all other cases the coefficient of F is defined with the aid of formulas of shorter length. In case b1 the term $\lambda x_1\ldots x_n A(x_1,\ldots,x_n)$ belongs to the same class as F, so it is not replaced as a whole. Thus the replacement can always be interpreted as taking place in shorter formulas. Hence: a) A shorter formula of a smaller class than σ has no significance for the coefficient of F. b) For a shorter formula of the class σ the replacement will not change the class and coefficient number, namely by the strengthening of proposition 46 and the induction hypothesis respectively. It follows that F* has the same coefficient as F.

For every formula that is not an atomic formula we define the following <u>preformulas</u>:

1. A formula $(t_1,\ldots,t_n \in \lambda x_1\ldots x_n A(x_1,\ldots,x_n))$ has the preformula $A(t_1,\ldots,t_n)$.

2. A formula $\neg A$ has the preformula A.

3. The formulas $(A \wedge B)$, $(A \vee B)$, and $(A \supset B)$ have the preformulas A and B.

4. The formulas $(\forall x)A(x)$ and $(\exists x)A(x)$ with a bound variable x of the type τ have as preformulas all the $A(t)$ in which t is a term of a type $\leq \tau$.

Proposition 48. Every preformula V of a formula F has one of the two properties:

a) V belongs to a smaller class than F, or

b) V belongs to the same class as F but has a smaller coefficient than F.

Proof. We must consider four different cases for F.

Case 1. Let F be a formula $(t_1,\dots,t_n \in \lambda x_1\dots x_n A(x_1,\dots,x_n))$ of the class σ with bound variables $x_1, \dots, x_n$ of the types $\tau_1, \dots, \tau_n$. Let $x_1, \dots, x_n$ be variables of these types. Every term t_i has a type $\leq \tau_i$, for otherwise F would not be a formula. In this case the preformula V is $A(t_1,\dots,t_n)$. By proposition 46 the class of the formula V is at most equal to the class σ_0 of the formula $A(x_1,\dots,x_n)$. By the definition rules t3 and F1, $\sigma_0 \leq \sigma$. Thus the class of V is at most equal to the class of F. If V and F belong to the same class, $\sigma_0 = \sigma$. Then the variables $x_1, \dots, x_n$. whose classes are in any case smaller than σ, belong to smaller classes than the formula $A(x_1,\dots,x_n)$. By proposition 47, V has the same coefficient as the formula $A(x_1,\dots,x_n)$. By b1 the formula F has a larger coefficient than $A(x_1,\dots,x_n)$ in the case $\sigma_0 = \sigma$. One thus obtains a coefficient for V smaller than that of F when V and F belong to the same class.

Case 2. Let F be a formula $\neg A$.

Then V is the formula A. This formula belongs to the same class as F and has a smaller coefficient than F.

Case 3. Let F be a formula $A \wedge B$, $A \vee B$, or $A \supset B$.

Then V is either A or B. Here the assertion of the propositions holds by the definition b3 of the coefficients.

Case 4. Let F be the formula $(\forall x)A(x)$ or $(\exists x)A(x)$ of the class σ with a variable x of type τ. The preformula V is a formula $A(t)$, whose class is at most equal to the class σ_0 of the formula $A(x)$ by proposition 46. By the definition F4, $\sigma_0 \leq \sigma$. The rest of the assertion is proved as in case 1.

We will now prove some syntactic properties of the system T. We say a signed formula has the degree, class, coefficient, etc., of its body. We define the rank of a formula or derivation as in Ω.

Proposition 49. Let X be a formula of the class σ with a coefficient b. Then $\{X,\overline{X}\}$ has a closed tableau in T of rank $\leq \omega.\sigma + 3\cdot b + 2$.

Proof. We show this by a simultaneous induction on the class and coefficient of X.

Let the class and coefficient of X be 1 and 0 respectively, i.e., the minimal possibilities. By considering the definition of a coefficient, one can see that in this case X must be an atomic formula. But then

$$X$$
$$\overline{X}$$

is a closed tableau for $\{X,\overline{X}\}$ of rank $\leq \omega.\sigma + 2$ (it is even of rank ≤ 0).

Let X not be atomic. Then the class σ and the coefficient b are not both minimal. We take as an induction hypothesis that the proposition holds for any set $\{Y,\overline{Y}\}$, where Y either belongs to a class $< \sigma$ or has a coefficient $< b$. By proposition 48 this holds for every preformula Y of X. For the class σ_0 and coefficient b_0 of such a preformula Y one obtains the following bounds:

a) if $\sigma_0 < \sigma$,

$$\omega.\sigma_0 + 3\cdot b_0 + 2 < \omega.\sigma_0 + \omega = \omega.(\sigma_0 + 1) \leq \omega.\sigma;$$

b) If $\sigma_0 = \sigma$ and $b_0 < b$,

$$\omega.\sigma_0 + 3\cdot b_0 + 2 < \omega.\sigma + 3\cdot b.$$

In either case, then, $\omega.\sigma_0 + 3\cdot b_0 + 2 < \omega.\sigma + 3\cdot b$ and hence by the induction hypothesis $\{Y,\overline{Y}\}$ has a closed tableau of rank $\leq \omega.\sigma + 3\cdot b$.

Thus if X is a γ^{τ} and $\overline{X}$ is consequently a δ^{τ}, for each term $t^{\tau'}$, $\tau' \leq \tau$, there is a closed tableau

$$\overline{X(t^{\tau'})}$$
$$X(t^{\tau'})$$
$$\mathcal{T}_{t\tau'}$$

for $\{X(t^{\tau'}), \overline{X(t^{\tau'})}\}$ of rank $\leq \omega \cdot \sigma + 3\cdot b$. But then

$$X$$
$$\overline{X}$$
$$\overline{X}(t^{\tau'})$$
$$X(t^{\tau'})$$
$$\mathcal{T}_{t\tau'}$$

is a closed tableau for $\{X,\overline{X}\}$ of rank $\leq \omega.\sigma + 3\cdot b + 2$.

The cases where X is an α (and $\overline{X}$ a β) and where X is a λ formula are completely analogous.

Since T contains the α and β rules and any branch of a tableau containing a formula and its conjugate can be extended to a closed branch (by proposition 49), any tautology can be proven in T, i.e., T is propositionally complete.

We can also show with a proof identical to that for Λ:

Theorem 37. The cut rule is eliminable in T.

As usual this yields:

Theorem 38. The system T is syntactically consistent, i.e., there is no formula A such that both A and $\overline{A}$ are provable in T.

§3. Systems of Analysis in Ramified Type Theory: Γ and Γ^*

In this paragraph we restrict ourselves to a part of our ramified type logic. Namely, we limit ourselves only to predicates all of whose argument places refer to the base domain. At the samy time we restrict ourselves to primitive constants and function symbols that were allowed in the number theoretic systems Ψ and Ω as predicate and function symbols. We thus make the following restrictions on the system T:

1. Besides the type 0, only the types of the form $\sigma(0,\ldots,0)$ will be used, where $0 < \sigma$.

2. As primitive constants we will use only the symbol 0 and n-place predicate symbols. The primitive constant 0 has the type 0. Every n-place predicate symbol will correspond to an evaluable predicate and will have the type $1(0,\ldots,0)$ with n zeros. (The equality symbol is assumed to be included among the two-place predicate symbols as in Ψ and Ω.)

3. All function symbols correspond to evaluable functions. (The apostrophe is assumed to occur among the one-place function symbols as the symbol for the successor function.)

The natural numbers are represented by terms of type 0. The terms of higher type, which we call predicates, represent sets of natural numbers or of n-tuples of natural numbers. We will consider systems with the restrictions 1 - 3 in which propositions about natural numbers and sets of natural numbers can be formalized. The sets will belong to different classes on the basis of the classes for predicates.

We first define such a system Γ in which the system Ψ of finitary number theory is contained. Here we call the variables (terms) of type 0 number variables (number terms) and the variables (terms) of the type $\sigma(0,\ldots,0)$ with n zeros n-place predicate variables (predicates) of the class σ. We use only finite classes in the system Γ.

The primitive symbols of the system Γ are:

a) The symbol 0, n-place primitive predicates (as symbols for evaluable predicates of the class 1) and n-place function symbols (as symbols for evaluable functions);

b) Number and predicate parameters and variables;

c) The operators $\neg, \wedge, \vee, \forall, \exists, \lambda$;

d) Parentheses and the comma.

(To every predicate parameter and variable there must correspond a finite place number $n \geq 1$ and a finite class number $\sigma \geq 1$.)

The definition of a <u>number terms</u> is:

t1. The symbol 0 and every number parameter and variable is a number term.

t2. If ρ is an n-place function symbol and $t_1, \ldots, t_n$ are number terms

$$\rho(t_1,\ldots,t_n)$$

is a number term.

We say a number term is a <u>numerical term</u> when it contains no parameters or variables.

Corresponding to the simultaneous definition of the terms and predicates of the system T we have here:

p1. Every n-place primitive predicate is an n-place predicate of the class 1.

p2. Every n-place predicate variable and predicate parameter of the class σ is an n-place predicate of the class σ.

p3. If $A(x_1,\ldots,x_n)$ is a formula of the class σ with number variables $x_1, \ldots, x_n$,

$$\lambda x_1\ldots x_n A(x_1,\ldots,x_n)$$

is an n-place predicate of the class σ.

F1. If p is an n-place predicate of the class σ and $t_1, \ldots, t_n$ are number terms,

$$pt_1,\dots,t_n$$

(usually written $t_1,\dots,t_n \in p$) is a formula of the class σ.

F2. If A is a formula of the class σ, $\neg A$ is also a formula of the class σ.

F3. If A and B are formulas, $(A \wedge B)$, $(A \vee B)$, and $(A \supset B)$ are also formulas. Their class is the maximum of the classes of A and B.

F4. If $A(x)$ is a formula of the class σ_0 with a number variable x (an n-place predicate variable x of the class σ_1), then

$$(\forall x)A(x) \quad \text{and} \quad (\exists x)A(x)$$

are also formulas. Their class is σ_0 (the maximum of σ_0 and σ_1).

A formula

$$(t_1,\dots,t_n \in p)$$

built by F1 in which p is a predicate built by p1 or p2 (a primitive predicate or a parameter) is called an <u>atomic formula</u>.

We will speak of closed terms and formulas from now on. We will call a formula or term <u>constant</u> if it has no parameters.

As in the system Ψ we call two formulas "equivalent" when one results from the other when particular numerical terms are replaced by numerical terms of the same numerical value. Similarly we define two formulas being "equivalent relative to $s = t$." We also define "arithmetically unsatisfiable" sets of formulas and "arithmetically true and false" formulas as in the system Ψ (of Chapter I, §1).

We now define the tableau system for Γ to be just like that of the system Ψ' except that the λ rule is added and certain type restrictions are made. Thus the rules for constructing a tableau for a set S in Γ are:

<u>α rule</u>

$$\begin{array}{c} \alpha \\ | \\ \alpha_1 \end{array} \qquad \begin{array}{c} \alpha \\ | \\ \alpha_2 \end{array}$$

<u>β rule</u>

$$\begin{array}{ccc} & \beta & \\ \swarrow & & \searrow \\ \beta_1 & & \beta_2 \end{array}$$

<u>γ rule</u>

$$\begin{array}{l} \gamma^{\tau} \\ \Big| \;,\ \tau' \leq \tau \\ \gamma^{\tau}(t^{\tau'}) \end{array}$$

<u>δ rule</u>

$$\begin{array}{l} \delta^{\tau} \\ \Big| \quad a^{\tau} \text{ new} \\ \delta^{\tau}(a^{\tau}) \end{array}$$

<u>λ rule</u>

$$\begin{array}{l} \pi t_1,\ldots,t_n \in \lambda x_1\ldots x_n A(x_1,\ldots,x_n) \\ \qquad | \\ \pi\ A(t_1,\ldots,t_n) \end{array}$$

<u>Cut rule</u>

$$\begin{array}{ccc} & X & \\ \swarrow & & \searrow \\ Y & & \overline{Y} \end{array}$$

<u>Assumption Rule</u>

$$\begin{array}{l} s_1 \\ \cdot \\ \cdot \qquad s_i \in S,\quad i = 1,\ldots,n \\ \cdot \\ s_n \end{array}$$

A branch θ in a tableau in Γ is <u>closed</u> if one of the following conditions holds:

a) θ contains an arithmetically unsatisfiable finite set of atomic formulas;

b) θ contains a set $\{X, \overline{X'}\}$, where X and X' are equivalent atomic formulas;

c) θ contains a set $\{Ts = t, X, \overline{X'}\}$ where X and X' are atomic formulas that are equivalent relative to $s = t$;

d) θ contains a formula $F(A(0) \wedge (\forall x)(A(x) \supset A(x'))) \supset A(t)$, in which x is a number variable and t is a number term.

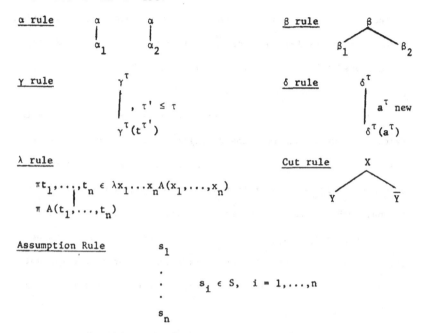

It is clear that the system Ψ' is contained in the system Γ. By proposition 18, Ψ is thus contained in Γ. We can prove by induction on the number of symbols in a formula that propositions like propositions 1-7 hold for Γ. In particular, since $\{X, \overline{X}\}$ has a closed tableau for any formula X and Γ has the α and β rules, Γ is propositionally complete.

Just as we went from the finitary system Ψ to the infinitary system Ω, we now go from the finitary system Γ to an infinitary system Γ^*. In Γ^* we do not restrict ourselves as we do in Γ to only finite classes, but rather allow as classes arbitrary ordinal numbers which are defined constructively in some manner.

The <u>primitive symbols</u> of the system Γ^* consist of the primitive symbols of Γ and of parameters and bound predicate variables of additional infinite classes. Here, however, parameters of type 0 are not used.

The number terms, predicates and formulas are defined as in Γ by the rules t1, t2, p1 - p3, and F1 - F4.

We note that we leave parameters in for types $\neq 0$. (In Ω we had no parameters at all.) This is because we will not use an infinitary δ rule for such types.

The <u>constant atomic formulas</u> of the system Γ^* have the form

$$(t_1,\ldots,t_n \in p)$$

with a primitive predicate p (of the class 1) and numerical terms $t_1, \ldots, t_n$. Since all primitive predicates are evaluable, every constant atomic formula has a truth value t or f. Let Γ_f^* be the class of constant atomic (signed) formulas πA such that

π is F if A is true and π is T if A is false.

The tableau system for Γ^* is now defined as follows. We have the rules (for a set S):

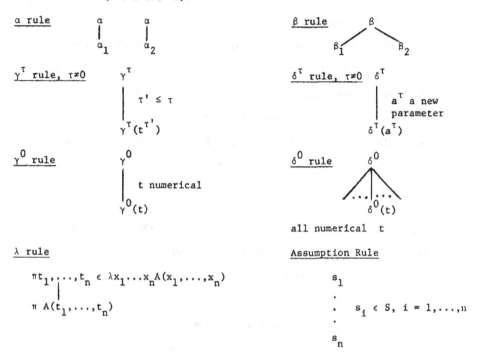

<u>α rule</u>

$$\frac{\alpha}{\alpha_1} \qquad \frac{\alpha}{\alpha_2}$$

<u>β rule</u>

$$\frac{\beta}{\beta_1 \mid \beta_2}$$

<u>γ^τ rule, $\tau \neq 0$</u>

$$\frac{\gamma^\tau}{\gamma^\tau(t^{\tau'})} \quad \tau' \leq \tau$$

<u>δ^τ rule, $\tau \neq 0$</u>

$$\frac{\delta^\tau}{\delta^\tau(a^\tau)} \quad a^\tau \text{ a new parameter}$$

<u>γ^0 rule</u>

$$\frac{\gamma^0}{\gamma^0(t)} \quad t \text{ numerical}$$

<u>δ^0 rule</u>

$$\frac{\delta^0}{\ldots \mid \delta^0(t) \mid \ldots}$$

all numerical t

<u>λ rule</u>

$$\frac{\pi t_1,\ldots,t_n \in \lambda x_1\ldots x_n A(x_1,\ldots,x_n)}{\pi\, A(t_1,\ldots,t_n)}$$

<u>Assumption Rule</u>

$$\begin{array}{l} s_1 \\ \cdot \\ \cdot \\ \cdot \\ s_n \end{array} \qquad s_i \in S,\ i = 1,\ldots,n$$

A branch in a tableau in Γ^* is <u>closed</u> if either:

a) it contains a formula of Γ^*_f; or

b) it contains a set $\{X, \overline{X'}\}$ of formulas such that X and X' are equivalent atomic formulas.

We note that we can prove the assertions of propositions 1-9 for Γ and Γ^* (by induction on the number of symbols in formula) just as for Ψ and Ω.

As with Ψ and Ω we will now see that Γ is a subsystem of Γ^* and that Γ^* is consistent because cuts are eliminable in it. For this we will need a few more definitions (we use the usual rank definition).

In Γ and Γ^* we say a <u>cut</u> is of a certain <u>class</u> or has a certain <u>coefficient</u> if its cut formula is in that class or has that coefficient. In Γ we say an <u>induction formula</u>

$$F(A(0) \wedge (\forall x)(A(x) \supset A(x'))) \supset A(t)$$

is of class σ and coefficient b if $A(0)$ is of class σ and coefficient b.

We can prove like theorem 5:

<u>Theorem 39</u>. If S is a set of formulas in which number parameters do not occur and $\mathcal{T}$ is a closed Γ tableau for S, τ can be transformed into a closed Γ^* tableau $\mathcal{T}'$ for S of rank $\leq \omega + \omega$, but perhaps with cuts. Moreover, let P be the set

$$\{\langle\sigma,b\rangle \mid \text{there is a cut or induction formula in } \mathcal{T} \text{ of class } \sigma \text{ and coefficient } b\}.$$

Then for every cut in $\mathcal{T}'$, there is a pair $\langle\sigma, b\rangle$ in P such that σ is the class of the cut and b is its coefficient.

Now we will show the consistency of Γ^* and hence of Γ by proving a cut elimination theorem similar to that for Ω. We will not make the restriction that cuts of only finitely many degrees (or classes or coefficients) occur in a tableau from which cuts are to be removed, although we will make other constructive restrictions. Thus within the context of the cut elimination theorem for Γ^* we will consider a tableau τ such that:

1) there is an ordinal μ such that the rank of $\mathcal{T}$ is $\leq \mu$;

2) there is an ordinal σ such that all cuts in $\mathcal{T}$ are of class $\leq \sigma$;

3) for any ordinal σ, there is (i.e., one can produce) an ordinal $b^{\sigma} \leq \omega$ such that every cut in $\mathcal{T}$ of class σ has a coefficient $\leq b^{\sigma}$.

We note that the Γ^* proofs that result from Γ proofs clearly satisfy these conditions.

Now say that the *class of a tableau* $\mathcal{T}$ is $\leq \sigma$ if for every σ_0 that is the class of a cut in $\mathcal{T}$, $\sigma_0 \leq \sigma$. The class of a $\mathcal{T}$ tableau is thus ≤ 0 iff no cut occurs in it. We say a tableau τ of class $\leq \sigma$ is of a class $< \sigma$ if there is an ordinal $\sigma' < \sigma$ such that $\mathcal{T}$ is of class $\leq \sigma'$. We say that a tableau $\mathcal{T}$ has a *σ-coefficient* $\leq b$ if every cut in $\mathcal{T}$ of class σ has a coefficient $\leq b$, where we restrict ourselves to σ-coefficients $\leq \omega$. We say a tableau $\mathcal{T}$ has a σ-coefficient $< b$ if there is a $b' < b$ such that $\mathcal{T}$ has a σ-coefficient $\leq b'$.

We see similarly to proposition 10:

Proposition 50. Let S be a set of formulas, $X(p)$ be a formula containing an n-place predicate parameter p of a class σ such that p does not occur in any formula of S, q be an n-place predicate of class $\leq \sigma$. Then if $S \cup \{X(p)\}$ has a closed tableau

$\mathcal{T}$ in Γ^* class $\leq \sigma$, σ-coefficient $\leq b$, and rank $\leq \mu$, where $\mu \geq \omega$, $\mathcal{T}$ can be transformed into a closed tableau for $S \cup \{X(q)\}$ of class $\leq \sigma$, σ-coefficient $\leq b$, and rank $\leq \mu$.

Proof. Assume $\mathcal{T}$ given satisfying the conditions of the proposition. As in proposition 10, we can transform $\mathcal{T}$ into a tableau $\mathcal{T}'$ for $S \cup \{X(q)\}$ in which none of the parameters in q are introduced into $\mathcal{T}'$ by δ^{τ} rules. Then if we substitute q for p throughout τ' we get a tableau $\mathcal{T}''$ in Γ^* for $S \cup \{X(q)\}$. Certain branches, however, may not be closed. Only in the case when a branch θ' in $\mathcal{T}'$ contains two equivalent formulas $\{T\ s_1,\ldots,s_n \in p,\ F\ t_1,\ldots,t_n \in p\}$ and q is not a parameter is the resulting branch in $\mathcal{T}''$ not closed. We have noted, however, that we have proposition 2 for Γ^* also. Thus we can extend such a θ'' to a closed branch by adding on only a tree of finite rank. We can actually see, moreover, that there is an integer m depending only on q such that all such trees have rank $\leq m$ (recall that here in Γ^* $s_1,\ldots,s_n$, $t_1,\ldots,t_n$ can only be <u>number terms</u>). Thus, by the rank lemma 2, the extension of the tableau $\mathcal{T}''$ has rank $\leq m + \mu$, which is equal to μ since $\omega \leq \mu$.

We show in a step by step manner similar to that we used for Ω that cuts can be eliminated in Γ^*.

<u>Proposition 51</u>. Every closed tableau with cuts of rank $\leq \mu$, class $\leq \sigma$, $\sigma \neq 0$, and σ-coefficient $b \leq n$, $n \neq \omega$, can be transformed into a closed tableau for the same set that has a rank $\leq \omega^{\mu}$ and either

belongs to a class $< \sigma$ or is of class σ but has a coefficient $< b$.

Proof. Let us be given a tableau

$$(*) \qquad \begin{array}{c} Y_1 \\ \cdot \\ \cdot \\ \cdot \\ Y_n \\ \mathcal{T} \end{array}$$

for a set S $(Y_1,\ldots,Y_n \in S)$ satisfying the assumptions of the proposition. Let us call a tableau "proper" if it satisfies the conclusion of the proposition, i.e., is of class $< \sigma$ or of class $\leq \sigma$ and of σ-coefficient $< b$. As an induction hypothesis we assume the theorem holds for all tableaux of rank $< \mu$. We now wish to transform (*) to a proper tableau.

If Y_n has no successors in (*), the proposition is trivially true. Otherwise let formulas X_i be the successors of Y_n in $\mathcal{T}$ and let the formula X_i have rank $\leq \mu_i$ $(< \mu)$. The formulas X_i are themselves at the tops of trees S_i. Let T_i be the tree

$$\begin{array}{c} Y_1 \\ \cdot \\ \cdot \\ \cdot \\ Y_n \\ S_i . \end{array}$$

It is easy to see that T_i can be viewed as a closed tableau for $S \cup \{X_i\}$ that is of rank $\leq \mu_i$. It is clearly also of class $\leq \sigma$ and has a σ-coefficient $\leq b$. By the induction hypothesis these trees can be transformed into closed proper tableaux T_i*:

$$\begin{array}{c} Y_1 \\ \cdot \\ \cdot \\ \cdot \\ Y_n \\ S_i^* \end{array}$$

for $S \cup \{X_i\}$ that are of rank $\leq \omega^{\mu}i$.

As in Ω we have various cases to consider.

1) Let the rule used to obtain the X_i not be a cut. Then we form a new tableau by putting the S_i* beneath the $Y_1,\ldots,Y_n$, using the same rule as in the original tableau to obtain the formulas X_i (which are at the top of the S_i*). By the definition of rank and properties of the ordinals (see the introduction), the rank of the resulting proper tableau for S is $\leq \omega^{\mu}$.

2) Let the rule used to obtain the X_i be a cut (so $i = 2$). This time we have to consider subcases depending on the kind of formulas the X_i are.

a. Let the X_i be atomic formulas. In such a case no rules can be applied to them. As in our proof for Ω, we form a new tree as follows. We take S_1*, delete X_1 from the top of it, and place the resulting subtree(s) below $Y_1,\ldots,Y_n$. We then take S_2*, delete X_2 from the top of it, and place the resulting subtrees below the end point of every branch of the tree we had just formed using S_1*. We claim the resulting tree closed is a proper tableau for S of rank $\leq \omega^{\mu}$. The argument that it is a closed tableau for S of rank $\leq \omega^{\mu}$ is the same as for Ω (cf. Chapter I, §4;

we must, however, replace the "2" in that proof by "ω"). And it is obvious that the tableau is proper. For if T_1^* and T_2^* are both of class $< \sigma$, (of class $\leq \sigma$ and of coefficients $b_1, b_2 < b$) the resulting tableau is clearly of class $< \sigma$ (of class $\leq \sigma$, but of σ-coefficient $< b$). And if one is of class $< \sigma$ and the other of class $\leq \sigma$ but of σ-coefficient $< b$, the resulting tableau is of class $\leq \sigma$ and of σ-coefficient $< b$.

b. Let X_1 be an α, so that X_2 is $\bar{\alpha}$, a β formula. This case is handled like the same case for Ω (in the proof of proposition 22). We need only note that in the tableau we form, we have besides the cuts that were in the T_i^* only cuts of the form

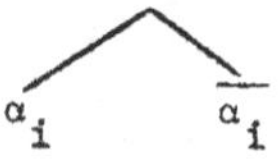

all of which are of the same class and coefficient. If that class is σ, the σ-coefficient must be $< b$, for otherwise the coefficient of

(which would also have class σ) would have to be $> b$, contrary to assumption. Since the class of α_i and $\bar{\alpha}_i$ is the same as that of α, it is clear then, that for these particular cuts either their class is $< \sigma$ or their class is $= \sigma$ and their σ-coefficient is $< b$. We can see as in the previous case that if we ignored these cuts, the tableau we formed would be of class $\leq \sigma$ and σ-coefficient $< b$. But then we see again by the same kind of argument as in the previous case that the same assertion holds when we take the new cuts into consideration. Thus the resulting tableau is proper.

c. Let X_1 be a γ^0 formula, so that X_2 is $\overline{\gamma^0}$, a δ^0 formula. This case is as the same case in the proof of proposition 22 with an additional argument about properness similar to the one given in b above.

d. Let X_1 be a γ^τ, $\tau \neq 0$ formula so that X_2 is $\overline{\gamma^\tau}$, a δ^τ formula. This case is handled very similarly to the case of a $\gamma^0 - \overline{\gamma^0}$ cut. Thus we have trees

$$T_1^*: \quad \begin{array}{c} Y_1 \\ \vdots \\ Y_n \\ \gamma^\tau \\ A \end{array} \qquad\qquad T_2^*: \quad \begin{array}{c} Y_1 \\ \vdots \\ Y_n \\ \overline{\gamma^\tau} \\ B \end{array}$$

where A and B are sets of trees. Now if no rule is ever applied to γ^τ $(\overline{\gamma^\tau})$ in T_1^* (T_2^*) then

$$\begin{array}{c} Y_1 \\ \vdots \\ Y_n \\ A \end{array} \qquad\qquad \begin{array}{c} Y_1 \\ \vdots \\ Y_n \\ B \end{array}$$

is a proper tableau of rank $\leq \omega^\mu$ for S.

Otherwise, first let a^τ be a parameter of type τ that does not occur in T_1^* or T_2^*. Let

$$\begin{array}{c} Y_1 \\ \vdots \\ Y_n \\ \overline{\gamma^\tau(a^\tau)} \\ B^{a^\tau} \end{array}$$

be the tableau that results from T_2^* when every occurrence of a $\overline{\gamma^\tau(c^\tau)}$ coming from $\overline{\gamma^\tau}$ is deleted and c^τ is replaced by a^τ in all its descendents. It is clear that the resulting tableau is a closed proper tableau for $S \cup \{\gamma^\tau(a^\tau)\}$ of rank $\leq \omega^{\mu 2}$ and of the same class σ' and σ'-coefficient as T_2^*. Let $t^{\tau'}$ be arbitrary predicate term of type $\leq \tau$, and let

$$\begin{array}{c} Y_1 \\ \vdots \\ Y_n \\ \hline \gamma^\tau(t^{\tau'}) \\ B^{t^{\tau'}} \end{array}$$

be the closed proper tableau of rank $\leq \omega^{\mu 2}$ for $S \cup \{\gamma^\tau(t^{\tau'})\}$ that we can find by proposition 50.

Now we form a tree

$$\begin{array}{c} Y_1 \\ \vdots \\ Y_n \\ C \end{array}$$

where C is the result of replacing each point $\gamma^\tau(t^{\tau'})$ $(\tau' \leq \tau)$ in A that is derived from γ^τ by a cut

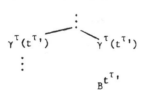

$$\begin{array}{ccc} & \vdots & \\ \gamma^\tau(t^{\tau'}) & & \gamma^\tau(t^{\tau'}) \\ \vdots & & \\ & & B^{t^{\tau'}} \end{array}$$

The rest of the argument for this case is just like that in case c.

e. Let X_1 and X_2 be λ formulas. This case is similar to (actually somewhat simpler than) case b.

This completes the proof of proposition 51.

<u>Proposition 52</u>. Every closed tableau with cuts of rank $\leq \mu$, class $\leq \sigma$, and σ-coefficient $b \leq n$, $n \neq \omega$, can be transformed into a closed tableau for the same set with rank $\leq \omega_{b+1}(\mu)$ and class $\leq \sigma$, but with no cut of class σ.

Proof. We show this by complete induction on b. When $b = 0$, the required tableau exists by proposition 51 with a rank $\leq \omega^{\mu} = \omega_{b+1}(\mu)$. If $b > 0$ by proposition 51 there is a closed tableau of rank $\leq \omega^{\mu}$ and either of class $< \sigma$ or of class $\leq \sigma$ and σ-coefficient $\leq b - 1$. In the first case we have a tableau satisfying our requirements. In the second case there is by the induction hypothesis a tableau of the required kind with rank $\leq \omega_b(\omega^{\mu}) = \omega_{b+1}(\mu)$.

We can now obtain the result of proposition 52 for tableaux of infinite σ-coefficients also.

<u>Proposition 53</u>. Every closed tableau with cuts of rank $\leq \mu$ and class $\leq \sigma$, $\sigma \neq 0$ can be transformed into a closed tableau for the same set with rank $\leq \epsilon_{\mu}$ and class $\leq \sigma$ which contains no cut of class σ.

Proof. Let there be given a closed tableau for a set S of rank $\leq \mu$ and class $\leq \sigma$, $\sigma \neq 0$:

$$(*) \qquad \begin{array}{l} Y_1 \\ \vdots \quad Y_i \in S,\ i = 1,\ldots,n \\ Y_n \\ \curlyvee \end{array}$$

In this proof we call a tableau "proper" if it is of a class $\leq \sigma$ but has no cut of class σ. We must transform (*) to a closed proper tableau for S of rank $\leq \epsilon_\mu$. We do this by transfinite induction on μ.

Let formulas X_i and tableaux of ranks $\leq \mu_i < \mu$ be defined as in proposition 51. By the induction hypothesis the tableaux T_i can be transformed into proper tableaux

$$(T_{i*}) \qquad \begin{array}{l} Y_1 \\ \vdots \\ Y_n \\ X_i \\ S_i{}^* \end{array}$$

for $S \cup \{X_i\}$ of rank $\leq \epsilon_{\mu_i}$.

If the X_i are not obtained by a cut of class σ we can form a proper tableau for S by placing the S_{i*} below

$$\begin{array}{c} Y_1 \\ \vdots \\ Y_n \\ \swarrow | \searrow \\ X_i \end{array}$$

The resulting tableau clearly has rank $\leq \epsilon_\mu$.

If the X_i are obtained by a cut of class σ, then

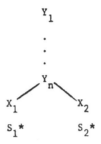

$$\begin{array}{ccc} & Y_1 & \\ & \vdots & \\ & Y_n & \\ X_1 & & X_2 \\ S_1^* & & S_2^* \end{array}$$

is a tableau of class $\leq \sigma$ with a single cut of class σ. It thus has a finite σ-coefficient b. It has rank $\leq \epsilon_\xi + 1$, where ξ is the maximum of μ_1 and μ_2. By proposition 52, this tableau can be transformed into a proper tableau for S of rank

$$\leq \omega_{b+1}(\epsilon_\xi + 1) < \epsilon_{\xi+1} \leq \epsilon_\mu .$$

If σ is a non-zero finite ordinal, and $\mathcal{T}$ is of class $\leq \sigma$, but has no cut of class σ, clearly $\mathcal{T}$ is also of a class $\leq \sigma - 1$. Thus it follows from proposition 53 that:

<u>Proposition 54</u>. Every closed tableau of class $\leq \sigma < \omega$ and of rank $\leq \mu$ can be transformed into a closed tableau for the same set of rank $\leq \epsilon_\mu$ and either of a class $< \sigma$ or without a cut (the former if $\sigma > 0$, the latter if $\sigma = 0$).

From proposition 54 we can obtain by complete induction on σ:

<u>Proposition 55</u>. Every closed tableau of a class $\leq \sigma < \omega$ and of rank $\leq \mu$ can be transformed into a <u>cut-free</u> tableau of rank $< \chi^\mu$ where χ^μ is any critical ϵ-number larger than μ.

Proof. Let us define $\epsilon^{\sigma}(\mu)$ as follows:

$$\epsilon_1(\mu) = \epsilon_{\mu}$$

$$\epsilon^{n+1}(\mu) = \epsilon_{\epsilon^n(\mu)}$$

Now we can prove by complete induction on σ (like proposition 52) that every closed tableau of a class $\leq \sigma < \omega$ and of rank $\leq \mu$ can be transformed into a cut-free tableau of rank $\leq \epsilon^{\sigma}(\mu)$. Now if $\mathcal{X}^{\mu}$ is a critical ϵ-number larger than μ,

$$\mu < \mathcal{X}^{\mu}$$

$$\epsilon_{\mu} < \epsilon_{\mathcal{X}^{\mu}} = \mathcal{X}^{\mu}$$

$$\epsilon_{\epsilon_{\mu}} < \epsilon_{\mathcal{X}^{\mu}} = \mathcal{X}^{\mu}$$

$$\vdots$$

$$\epsilon^{\sigma}(\mu) < \mathcal{X}^{\mu}$$

(cf. property 4a of the constructive ordinals in the appendix).

It is easy to see that if we have a tableau $\mathcal{T}$ of class $\leq \sigma$, either the tableau has a cut of class σ or we can find an ordinal $\sigma' \leq \sigma$ such that $\mathcal{T}$ is of class $\leq \sigma'$ and $\sigma' = 0$ or σ' is a limit ordinal. For if σ is not 0 or a limit ordinal we can find its predecessor and repeat this process until we find a σ' such that $\mathcal{T}$ is of class $\leq \sigma'$ and either $\mathcal{T}$ has a cut of class σ' or σ' is 0 or a limit ordinal. This process must end because there is no infinite decreasing sequence of ordinals. For our final theorem the tableaux we consider are assumed to have as classes ordinals which satisfy the property underlined above. In order to elimi- lined above. In order to eliminate cuts from tableaux of arbitrary class, we must define one more concept.

We say a tableau $\mathcal{T}$ of class $\leq \sigma$ has an <u>σ-index</u> σ if $\mathcal{T}$ contains no cut of class σ (in this case $\sigma = 0$ or σ is a limit ordinal). We say $\mathcal{T}$ has an σ-index $\sigma + 1$ if $\mathcal{T}$ contains at least one cut of the class σ. (We assume we know a class and index for any tableau we consider.) Finally we have:

<u>Theorem 40</u>. Every closed tableau of class $\leq \sigma$, σ-index $\leq \nu$ and rank $\leq \mu$ can be transformed into a cut-free closed tableau for the same set of rank $\leq \rho(\mu,\nu)$.

Proof. Here $\rho(\mu,\nu)$ is the function (mentioned in the introduction) such that:

(a) $\rho(\mu_i,\nu) < \rho(\mu,\nu)$ for $\mu_i < \mu$,

(b) $\rho(\mu,\nu_1) < \rho(\mu,\nu_2)$ for $\nu_1 < \nu_2$,

(c) $\rho(\epsilon_{\rho(\mu,\nu)+1},\nu_1) < \rho(\mu_1,\nu)$ for $\mu < \mu_1$ and $\nu_1 < \nu$.

Proof. Let there be given a closed tableau

$$\begin{array}{c} Y_1 \\ \cdot \\ \cdot \\ \cdot \\ Y_n \\ \mathcal{T} \end{array}$$

for a set S of class $\leq \sigma$, σ-index $\leq \nu$, and rank $\leq \mu$. We assume the assertion of the theorem holds for any closed tableau

(1) of class $\leq \sigma' \leq \sigma$, of σ-index $\leq \nu$ and rank $< \mu$, or

(2) of class $\leq \sigma' \leq \sigma$, of σ'-index $< \nu$ and any rank, or

(3) of class $< \sigma$.

If $\mathcal{T}$ is empty, or $\sigma = 0$ we are done. Otherwise let X_i be the successors of Y_n in $\mathcal{T}$ and let the rank of X_i be $\le \mu_i < \mu$. The X_i are at the top of the trees S_i and by the induction hypothesis we can obtain a cut-free tableau

$$\begin{array}{c} Y_1 \\ \cdot \\ \cdot \\ \cdot \\ Y_n \\ S_i^* \end{array}$$

for $S \cup \{X_i\}$ of rank $\le \rho(\mu_i, \nu)$.

If the X_i are not obtained by a cut in the original tableau

$$(*) \qquad \begin{array}{c} Y_1 \\ \cdot \\ \cdot \\ \cdot \\ Y_n \\ / \mid \backslash \\ S_i^* \end{array}$$

is a closed tableau for S of rank $\le \rho(\mu,\nu)$ by property a of ρ.

If the X_i are obtained by a cut of class $\sigma' < \sigma$, the tableau (*) has class $\sigma' < \sigma$. If $\nu = \sigma$, then σ is a limit number (we have assumed it is not 0) and the σ'-index ν_1 of (*) is $< \nu$. The σ'-index ν_1 of (*) is also clearly $< \nu$ if $\nu = \sigma + 1$. And by induction hypotheses 3, (*) can be transformed into a closed cut-free tableau for S. Since the rank of (*) is $\le \rho(\xi,\nu) + 1$, where ξ is the maximum of μ_1 and μ_2, the rank of (*) is also $< \epsilon_{\rho(\xi,\nu)+1}$. Then by the induction hypothesis the rank of the final tableau is $\le \rho(\epsilon_{\rho(\xi,\nu)+1}, \nu_1)$ which is $< \rho(\mu,\nu)$ by property (c) of ρ.

If the X_i are obtained by a cut of class σ, by proposition 53, one can obtain from (*) (which has rank $\leq \rho(\xi,\nu) + 1$, where ξ is the maximum of μ_1 and μ_2) a closed tableau for S with rank $\leq \epsilon_{\rho(\xi,\nu)+1}$ and class $\leq \sigma$, but with no cut of class σ. We then find an ordinal $\sigma' \leq \sigma$ such that $\sigma' = 0$ or σ' is a limit ordinal and the tableau is of class $\leq \sigma'$ but has no cut of class σ'. The σ'-index ν_1 of the resulting tableau is clearly $< \nu$. The argument about the rank is carried out as in the previous case.

We obtain from this as usual:

Theorem 41. There is no formula A such that both A and $\neg A$ are provable in Γ^*.

(Actually our discussion allows us to state only "there is no formula A such that both A and $\neg A$ are provable with tableaux satisfying certain constructive conditions. Since we use this result to obtain the consistency of Γ and all Γ^* tableaux that come from Γ tableaux satisfy the constructive conditions, our result is sufficient for our purposes.)

Using proposition 55, we can now strengthen theorem 39 as follows:

Theorem 42. If S is a set of formulas in which number parameters do not occur and $\mathcal{T}$ is a closed Γ tableau for S, $\mathcal{T}$ can be transformed into a closed cut-free Γ^* tableau for S of rank $\leq \varkappa_0$ (where $\varkappa_0$ is the first critical ϵ-number).

§4. Transfinite Induction in Γ and Γ^*

We will now investigate the derivations of formulas expressing transfinite induction in the systems Γ and Γ^*. We will as before consider the formula

$$J_x(A(x),t)$$

introduced in chapter I, §5. The induction proofs of the system Ψ given there also hold in the system Γ. Other derivations may be formed in Γ, however, We abbreviate

$$J_x((x \in q),t) \quad \text{by} \quad J(q,t)$$

when q is a one-place predicate, x is a bound number variable and t is a number term. With p^σ and x^σ we denote in this paragraph a one-place predicate parameter or variable of the class σ. We define

$$J_\sigma(t) = (\forall x^\sigma)J(x^\sigma,t)$$

for a number term t. This formula belongs to the class $\sigma + 1$. It expresses transfinite induction up to t in this class.

The iterated exponential function will be represented by a function ζ such that:

(1) $\zeta(0,t) = t$

(2) $\zeta(a',t) = E(\zeta(a,t))$.

By properties 11 and 12 of the ordinals, there is for every numeral z a natural number n with

$$z < \zeta(n,L(z) + 1).$$

Here $L(z)$ is the ordinal ω or the largest ϵ-number $\le z$, depending on whether $z < \epsilon_0$ or $\epsilon_0 \le z$ respectively. Thus there is a one-place evaluable function η such that in general

$$z < \zeta(\eta(z),\ L(z) + 1)$$

Since $L(z) = \omega$ for $z < \epsilon_0$ the formula

(3) $a < \epsilon_0 \supset a < \zeta(\eta(a),\ \omega + 1)$

is arithmetically true.

There is a one-place evaluable function τ with the properties

$$\tau(z) = 0 \quad \text{when} \quad z < \epsilon_0$$

$$\epsilon_{\tau(z)} = L(z) \quad \text{when} \quad \epsilon_0 \leq z.$$

In every case, $L(z) \leq \epsilon_{\tau(z)}$, so

$$z < \zeta(\eta(z),\ \epsilon_{\tau(z)} + 1),$$

Thus

(4) $b < \zeta(\eta(b),\ \epsilon_{\tau(b)} + 1)$

is arithmetically true.

If $z_1 < \epsilon_z$, then also $L(z_1) < \epsilon_z$. When $\epsilon_0 \leq z_1$, it follows that $\epsilon_{\tau(z_1)} = L(z_1) < \epsilon_z$, so that $\tau(z_1) < z$. When $z_1 < \epsilon_0$, $\tau(z_1) = 0$, so that either $\tau(z_1) < z$ or $z = 0$. Hence the formula

(5) $b < \epsilon_a \supset (\tau(b) < a \vee a = 0)$

is arithmetically true.

We use $K_\sigma(t)$ to denote $J_\sigma(\epsilon_t + 1)$. This formula has class $\sigma + 1$. We use $K^*_\sigma(t)$ to denote

$$(\forall x)(x < t \supset K_\sigma(x)).$$

(Here x is a number variable). It is easy to see that the following are provable formulas:

(6) $K^*_\sigma(a) \supset (\tau(b) < a \supset K_\sigma(\tau(b)))$

$J_{\sigma+1}(t) \supset J(p^{\sigma+1},\ t)$

$J_\sigma(t) \supset J(p^\sigma,\ t)$

$a = 0 \supset (J_\sigma(\epsilon_0) \equiv J_\sigma(\epsilon_a))$

One has further

$$J_{\sigma+1}(t) \supset J(\lambda x K_\sigma(x), t)$$
$$J_\sigma(t) \supset J(\lambda x A_\sigma(x), t),$$

where $A_\sigma(t)$ is an arbitrary formula of the class σ. Recalling that $J(\lambda x K_\sigma(x), t)$ is an abbreviation for $J_x((x \in \lambda x K_\sigma(x)), t)$ and similarly for the other formula, it is easy to see that

$$(7) \quad J_{\sigma+1}(t) \supset J_x(K_\sigma(x), t)$$
$$J_\sigma(t) \supset J_x(A_\sigma(x), t)$$

are also provable. We recall that these formulas are abbreviations for

$$(8) \quad J_{\sigma+1}(t) \supset (P_x K_\sigma(x) \supset K_\sigma^*(t))$$

and

$$J_\sigma(t) \supset (P_x A_\sigma(x) \supset A_\sigma^*(t)).$$

The formula

$$A_\sigma^*(t) \supset (a < t \supset A_\sigma(a))$$

is clearly a provable formula. It is easy to see with the use of cuts and the last two formulas that

$$(9) \quad J_\sigma(t) \supset (P_x A_\sigma(x) \supset (a < t \supset A_\sigma(a)))$$

is provable.

The fundamental formulas we will need to prove our induction formulas in Γ are the following, which we will prove in order:

I. $J_\sigma(a) \supset (b < a \supset J_\sigma(b))$,

II. $J_\sigma(a) \supset J_\sigma(a + 1)$,

III. $J_\sigma(\omega)$,

IV. $J_\sigma(t) \supset J_\sigma(E(t))$,

V. $J_\sigma(t) \supset J_\sigma(\zeta(s, t))$,

VI. $J_\sigma(\epsilon_0)$,

VII. $J_{\sigma+1}(a) \supset J_\sigma(\epsilon_a)$.

We will use the same conventions about the presentations of our proofs as in Chapter I, §5, for Ψ.

<u>Proof of proposition I.</u> We first note that

(10) $J(p^{\sigma},a) \supset (b < a \supset J(p^{\sigma},b))$

is derivable in Γ just as the "same" formula was derivable in Ψ. The formula I now has the following proof:

1 $F\ J_{\sigma}(a) \supset (b < a \supset J_{\sigma}(b))$

| α; 1

2 $T\ J_{\sigma}(a)$

| α; 1

3 $F\ b < a \supset J_{\sigma}(b)$

| α; 3

4 $T\ b < a$

| α; 3

5 $F\ J_{\sigma}(b)$

| AB; 2

6 $T(\forall x)J(x^{\sigma}, t)$

| AB; 5

7 $F(\forall x^{\sigma})J(x^{\sigma},b)$

| δ^{σ}; 7

8 $F\ J(p^{\sigma}, b)$

| γ^{σ}; 6

9 $T\ J(p^{\sigma}, a)$

| TH; (10)

10 $T\ J(p^{\sigma},a) \supset (b < a \supset J(p^{\sigma}, b))$

| β; 10

11 $F\ J(p^{\sigma},a)$ | (right branch)

$\$_b$; 9,11   (1)

(1)

12 $\quad T\ b < a \supset J(p^{\sigma},b)$

$\beta;\ 12$

13 $\quad F\ b < a$

$\$_b;\ 4,13$

14 $\quad T\ J(p^{\sigma},b)$

$\$_b;\ 8,14$

Now the formulas II and III can be proven like I using the fact that

$$J(p^{\sigma},a) \supset J(p^{\sigma},\ a + 1)$$

and

$$J(p^{\sigma},\ \omega)$$

are known to be provable from our proofs in Ψ.

Formula V of the Chapter I, §5 (for Ψ) shows that for any parameter p^{σ}, there is a formula $A_{\sigma}(a)$ of class σ for which

(11) $\quad J_x(A_{\sigma}(x),t) \supset J(p_{\sigma},\ E(t))$

is provable. Formula IV then has the proof:

1 $\quad F\ J_{\sigma}(t) \supset J_{\sigma}(E(t))$

$\alpha;\ 1$

2 $\quad T\ J_{\sigma}(t)$

$\alpha;\ 1$

3 $\quad F\ J_{\sigma}(E(t))$

$AB;\ 3$

4 $\quad F(\forall x^{\sigma})J(x^{\sigma},E(t))$

$\delta^{\sigma};\ 4$

5 $\quad F\ J(p^{\sigma},E(t))$

$TH;\ (11)$

(1)

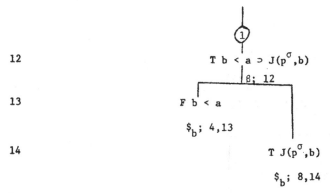

(1)

6 $T\ J_x(A_\sigma(x),t) \supset J(p^\sigma,E(t))$

TH; (7)

7 $T\ J_\sigma(t) \supset J_x(A_\sigma(x),t)$

β; 7

8 $F\ J_\sigma(t)$

$\$_b$; 2,8

9 $T\ J_x(A_\sigma(x),t)$

β; 6

10 $F\ J_x(A_\sigma(x),t)$

$\$_b$; 9,10

11 $T\ J(p^\sigma,E(t))$

$\$_b$; 5,11

Formula V has the following proof:

1 $F\ J_\sigma(t) \supset J_\sigma(\zeta(s,t))$

α; 1

2 $T\ J_\sigma(t)$

α; 1

3 $F\ J_\sigma(\zeta(s,t))$

CI;

4 $T(J_\sigma(\zeta(0,t)) \wedge (\forall x)(J_\sigma(\zeta(x,t) \supset J_\sigma(\zeta(x',t)))) \supset J_\sigma(\zeta(s,t))$

β; 4

5 $T\ J_\sigma(\zeta(s,t))$

$\$_b$; 3,5

6 $F\ J_\sigma(\zeta(0,t)) \wedge (\forall x)(J_\sigma(\zeta(x,t) \supset J_\sigma(\zeta(x',t)))$

β; 6

(1)

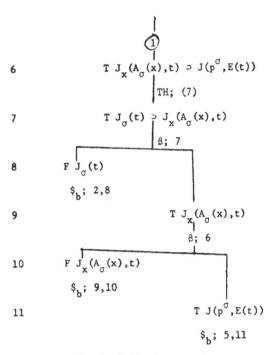

(1)

7 $J_\sigma(\zeta(0,t))$

AX; (1)

8 $T\ \zeta(0,t) = t$

$\$_c$; 3, 7, 8

9 $F(\forall x)(J_\sigma(\zeta(x,t)) \supset J_\sigma(\zeta(x',t)))$

δ; 9

10 $F\ J_\sigma(\zeta(a,t)) \supset J_\sigma(\zeta(a',t))$

TH; IV

11 $T\ J_\sigma(\zeta(a,t) \supset J_\sigma(E(\zeta(a,t))$

A; (2)

12 $T\ \zeta(a',t) = E(\zeta(a,t))$

$\$_c$; 10,11,12

Formula VI has the following proof in Γ:

1 $F\ J_\sigma(\epsilon_0)$

AB; 1

2 $F(\forall x^\sigma)\ J(x^\sigma,\epsilon_0)$

δ^σ; 2

3 $F\ J(p^\sigma,\epsilon_0)$

AB; 3

4 $F\ P_x(p^\sigma x) \supset (\forall x)(x < \epsilon_0 \supset p^\sigma x)$

α; 4

5 $T\ P_x(p^\sigma x)$

α; 4

6 $F(\forall x)(x < \epsilon_0 \supset p^\sigma x)$

δ; 6

7 $F\ a < \epsilon_0 \supset p^\sigma a$

TH; (3)

(1)

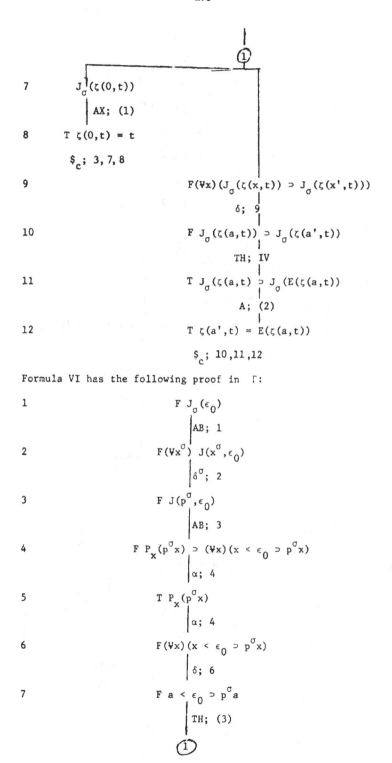

(1)

8 $T\ a < \epsilon_0 \supset a < \zeta(\eta(a), \omega + 1)$

TH; (9)

9 $T\ J_\sigma(\zeta(\eta(a), \omega + 1)) \supset (P_x(p^\sigma x) \supset (a < \zeta(\eta(a), \omega + 1) \supset p^\sigma a))$

α; 7

10 $T\ a < \epsilon_0$

α; 7

11 $F\ p^\sigma\ a$

β; 8

12 $F\ a < \epsilon_0$

$\#_a$; 10,12

13 $T\ a < \zeta(\eta(a), \omega + 1)$

TH; II

14 $T\ J_\sigma(\omega) \supset J_\sigma(\omega + 1)$

TH; III

15 $T\ J_\sigma(\omega)$

β; 14

16 $F\ J_\sigma(\omega)$

$\$_b$; 15,16

17 $T\ J_\sigma(\omega + 1)$

TH; V

18 $T\ J_\sigma(\omega + 1) \supset J_\sigma(\zeta(\eta(a), \omega + 1))$

β; 18

19 $F\ J_\sigma(\omega + 1)$

$\$_b$; 17,19

20 $T\ J_\sigma(\zeta(\eta(a), \omega + 1))$

(2)

(2)

β; 9

21 F $J_\sigma(\zeta(\eta(a),\ \omega + 1))$

$\$_b$; 20,21

22 T $P_x(p^\sigma x) \supset (a < \zeta(\eta(a),\ \omega + 1) \supset p^\sigma a$

β; 22

23 F $P_x(p^\sigma x)$

$\$_b$; 5, 23

24 T $a < \zeta(\eta(a),\ \omega + 1) \supset p^\sigma a$

β; 24

25 F $a < \zeta(\eta(a),\ \omega + 1)$

$\#_a$; 13,25

26 T $p^\sigma a$

$\$_b$; 11,26

We now prove formula VII. We will denote the part of the following closed tableau for $\{T\ K^*_\sigma(a),\ F\ J_\sigma(\epsilon_a)\}$ below the two initial formulas by $\mathcal{T}^{a}$:

1 T $K^*_\sigma(a)$

2 F $J_\sigma(\epsilon_a)$

AX; (6)

3 T $a = 0 \supset (J_\sigma(\epsilon_0) \supset J_\sigma(\epsilon_a))$

β; 3

4 T $J_\sigma(\epsilon_0) \supset J_\sigma(\epsilon_a)$

β; 4

5 F $J_\sigma(\epsilon_0)$

$\$_d$(theorem VI)

(1)

(2)

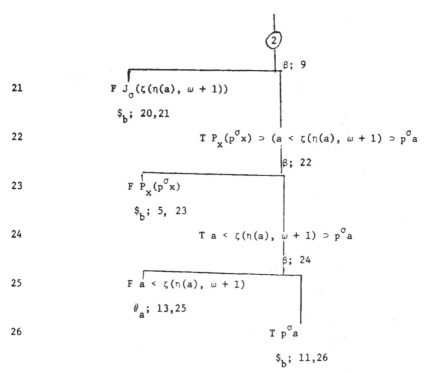

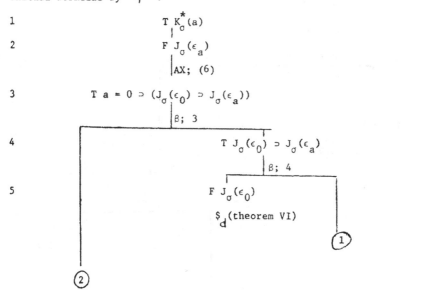

(2) | (1)

6 | $T\ J_{\sigma}(\epsilon_a)$

$\$_b$; 2,6

7 $F\ a = 0$

AB; 2

8 $F(\forall x^{\sigma})\ J(x^{\sigma}, \epsilon_a)$

δ^{σ}; 8

9 $F\ J(p^{\sigma}, \epsilon_a)$

AB; 9

10 $F\ P_x(p^{\sigma}x) \supset (\forall x)(x < \epsilon_a \supset p^{\sigma}x)$

α; 10

11 $T\ P_x(p^{\sigma}x)$

α; 10

12 $F(\forall x)(x < \epsilon_a \supset p^{\sigma}x)$

δ; 12

13 $F\ b < \epsilon_a \supset p^{\sigma}b$

AX; (5)

14 $T\ b < \epsilon_a \supset (\tau(b) < a \vee a = 0)$

β; 14

15 $F\ b < \epsilon_a$

α; 13

16 $T\ b < \epsilon_a$

$\#_a$; 15,16

17 $T\ \tau(b) < a \vee a = 0$

β; 17

18 $T\ a = 0$

$\#_a$; 7,18

19 $T\ \tau(b) < a$

(3)

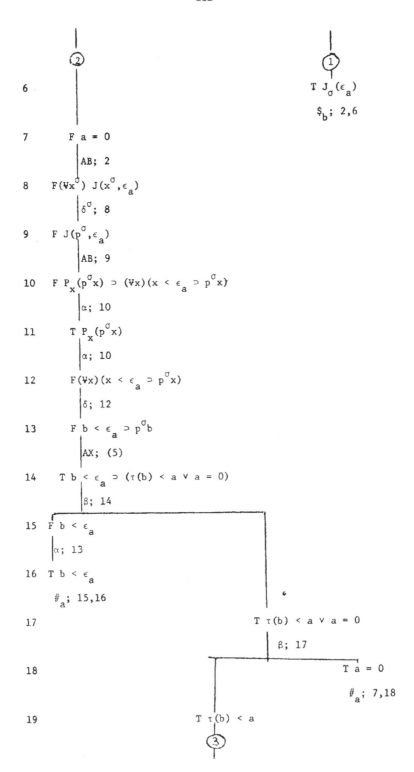

(3)

AX; (6)

20 $T\ K^*_\sigma(a) \supset (\tau(b) < a \supset K_\sigma(\tau(b))$

β; 20

21 $F\ K^*_\sigma(a)$

$\$_b$; 1,21

22 $T\ \tau(b) < a \supset K_\sigma(\tau(b))$

β;22

23 $F\ \tau(b) < a$

$\#_a$; 19,23

24 $T\ K_\sigma(\tau(b))$

AB; 24

25 $T\ J_\sigma(\epsilon_{\tau(b)} + 1)$

TH; V

26 $T\ J_\sigma(\epsilon_{\tau(b)} + 1) \supset J_\sigma(\zeta(\eta(b),\ \epsilon_{\tau(b)} + 1))$

β; 26

27 $F\ J_\sigma(\epsilon_{\tau(b)} + 1)$

$\$_b$; 25, 27

28 $T\ J_\sigma(\zeta(\eta(b),\ \epsilon_{\tau(b)} + 1))$

AX; (9)

29 $T\ J_\sigma(\zeta(\eta(b),\ \epsilon_{\tau(b)} + 1)) \supset (P_x(p^\sigma x) \supset (b < \zeta(\eta(b), \epsilon_{\tau(b)} + 1) \supset p^\sigma b))$

β; 29

30 $F\ J_\sigma(\zeta(\eta(b),\ \epsilon_{\tau(b)} + 1))$

$\$_b$; 28,30

31 $T\ P_x(p^\sigma x) \supset (b < \zeta(\eta(b),\ \epsilon_{\tau(b)} + 1) \supset p^\sigma b)$

(4)

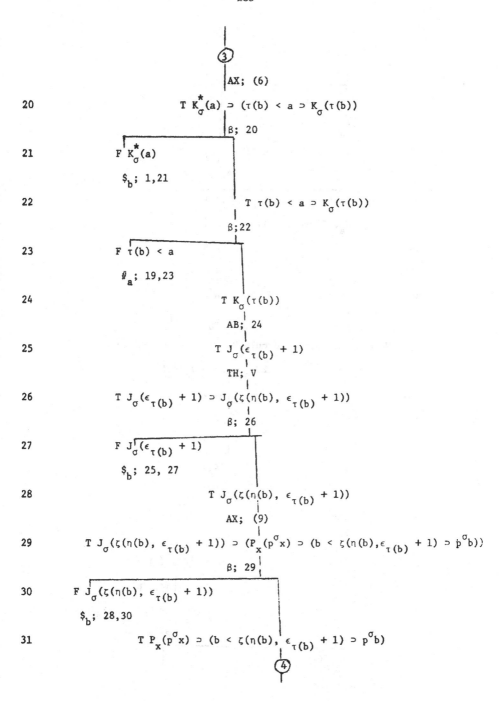

(4)

β; 31

32 F $P_x(p^\sigma x)$

$\$_b$; 11,32

33 T $b < \zeta(\eta(b), \epsilon_{\tau(b)} + 1) \supset p^\sigma b$

β; 33

34 F $b < \zeta(\eta(b), \epsilon_{\tau(b)} + 1)$

$\#_a$; 34 (cf. (4))

35 T $p^\sigma b$

α; 13

36 F $p^\sigma b$

$\$_b$; 35,36

We can then prove formula VII in Γ as follows:

1 F $J_{\sigma+1}(a) \supset J_\sigma(\epsilon_a)$

α; 1

2 T $J_{\sigma+1}(a)$

α; 1

3 F $J_\sigma(\epsilon_a)$

CUT

4 T $K^*_\sigma(a)$

γ^a

5 F $K^*_\sigma(a)$

AX; (8)

6 T $J_{\sigma+1}(a) \supset (P_x K_\sigma(x) \supset K^*_\sigma(a))$

β; 7

7 F $J_{\sigma+1}(a)$

$\$_b$; 2,8

(1)

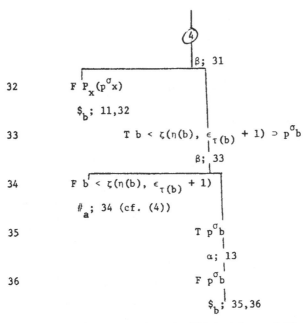

9 $T\ P_x K_\sigma(x) \supset K^*_\sigma(a)$

β; 9

10 $T\ K^*_\sigma(a)$

$\times^{a}$

11 $F\ P_x K_\sigma(x)$

AB; 11

12 $F(\forall x)(K^*_\sigma(x) \supset K_\sigma(x)$

δ; 12

13 $F\ K^*_\sigma(c) \supset K_\sigma(c)$

α; 13

14 $T\ K^*_\sigma(c)$

α; 13

15 $F\ K_\sigma(c)$

AB; 15

16 $F\ J_\sigma(\epsilon_c + 1)$

TH; II

17 $T\ J_\sigma(\epsilon_c) \supset J_\sigma(\epsilon_c + 1)$

β; 17

18 $F\ J_\sigma(\epsilon_c)$

$\times^{c}$

19 $T\ J_\sigma(\epsilon_c + 1)$

$\$_b$; 16,19

We say that transfinite induction is provable in Γ up to an ordinal, (i.e., a numeral considered as an ordinal) z when the the formula

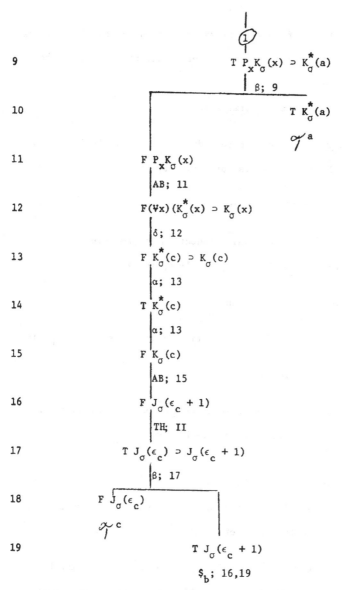

$$J_\sigma(z)$$

is provable in Γ for every class σ. (As in Chapter I, §5 we can generalize such an assertion to order relations other than the particular one we are assuming.)

One can prove using formula I:

<u>Proposition 56</u>. If transfinite induction is provable in Γ up to an ordinal z, it is also provable for every ordinal $z_1 < z$.

With formula II we obtain:

<u>Proposition 57</u>. If transfinite induction is provable in Γ up to an ordinal z, it is also provable up to $z + 1$.

From formula VI we obtain:

<u>Proposition 58</u>. Transfinite induction is provable up to ϵ_0 in Γ.

Formula VII gives us:

<u>Proposition 59</u>. If transfinite induction is provable up to an ordinal z in Γ, it is also provable up to ϵ_z.

Finally we obtain:

<u>Theorem 43</u>. Transfinite induction is provable in Γ up to every ordinal less than the first critical ϵ-number χ_0.

Proof. Let z be $< \chi_0$. We take as an induction hypothesis that transfinite induction is provable in Γ up to every ordinal $z_1 < z$.

If z is $\leq \epsilon_0$, the assertion holds by propositions 58 and 56. So let z be $> \epsilon_0$. Then $L(z) = \epsilon_{\tau(z)}$ and $\epsilon_{\tau(z)} \leq z < \epsilon_{\tau(z)+1}$. Since $z < \mathcal{X}_0$, we have $\tau(z) < \epsilon_{\tau(z)}$. Hence $\tau_{(z)} < z$, so that transfinite induction is provable in Γ up to $\tau(z)$ by the induction hypothesis. Then by proposition 57, it is also provable up to $\tau(z) + 1$ and by proposition 59 also up to $\epsilon_{\tau(z)+1}$. Since $z < \epsilon_{\tau(z)+1}$, by proposition 56 it is provable up to z.

We cannot, however, prove transfinite induction in Γ up to any ordinal z such that $\mathcal{X}_0 \leq z$. To see this we proceed as we did in the case of Ψ. We go from Γ^* to a system $\overline{\Gamma^*}$ by adding a primitive symbol $\bar{p}$ (we assume the formulas $\bar{p}t$ belong to class 1).

Corresponding to proposition 27 we have:

<u>Proposition 60</u>. If $\bar{p}z$ has a cut-free proof in $\overline{\Gamma^*}$ of rank $\leq r$, then $r \geq z$.

With this one proves like theorem 12:

<u>Theorem 44</u>. If $z \geq \mathcal{X}_0$, p^1 is of class 1, and $J_x(p^1x, z)$ has a cut-free proof in Γ^* of rank $\leq r$, then $r \geq \mathcal{X}_0$.

From this we obtain:

<u>Theorem 45</u>. A formula $J_\sigma(z)$ is provable in Γ iff $z < \mathcal{X}_0$.

Proof. If $z < \mathcal{X}_0$, $J_\sigma(z)$ is provable in Γ by theorem 44. If $J_\sigma(z)$ is provable in Γ, then $J_x(p^\sigma x, z)$ is provable in Γ

for an arbitrary parameter p of class σ, and by proposition 50 $J_x(p^4x,z)$ is provable in Γ for any parameter p^1 of class 1. But then by theorem 42 $J_x(p^1x,z)$ has a cut-free proof in Γ^* of rank $< \mathcal{X}_0$. By theorem 44, then, $z < \mathcal{X}_0$.

In the same way we extended the system Ψ to systems Ψ_α, we extend Γ to systems Γ_α by allowing the formulas $F\ J_\sigma(\alpha)$ to close a branch for every finite class σ. We can show as in Chapter I, §5:

Theorem 45. A formula $J_\sigma(z)$ is provable in Γ_α iff z is less than the first critical ϵ-number following α.

§5. The Interpretation of Analysis

In Chapter XI of Beweistheorie Schütte shows how substantial parts of classical analysis can be formalized within systems like Λ and T. The material is quite straightforward and in this case the reader is referred to his work via appendix I. As an example, in T a term p of class σ is said to be a real number of class σ if

$$(\forall x)(px \sim (\exists y)(py \wedge x \mathrel{\dot{<}} y))$$

is provable (The Dedekind cut characterization), where $\dot{<}$ is the inequality relation for the rational numbers (and $A \sim B$ is an abbreviation for $(A \supset B) \wedge (B \supset A)$). One just drops the class restrictions to obtain definitions in Λ. The fundamental classical proofs have to be revised somewhat, and Schütte exemplifies a general procedure that can be used with a proof that any function that is continuous in a proper closed interval is uniformly continuous on that interval.

It turns out that the type-free system is more general in that the ramified type theory can be represented within it. It is an open question, however, as to whether the number concept of type-free analysis is essentially broader than that of ramified analysis for in the most important cases in which one uses a proof that a particular term p is a real number in type-free analysis, one can also find a corresponding proof in ramified analysis.

APPENDIX 1 - Cut Elimination in First Order Logic as Repetition Introduction

The known constructive proofs of cut-elimination (for systems of various orders and kinds) tell one how to construct from a proof with cuts a possibly larger proof without cuts. The first question one might ask is whether or not these proofs really do have to grow sometimes, or, in other words, whether or not there is some theorem that has a proof with cuts that is shorter than any proof of the theorem without cuts. This question is rather easily answered by example already for propositional logic. For instance consider the following proof of $[(P\wedge A)\vee(P\wedge B)]\supset[(P\vee A)\wedge(P\vee B)]$ containing only one cut:

$$F\ [(P\wedge A)\vee(P\wedge B)]\supset[(P\vee A)\wedge(P\vee B)]$$

$$T\ (P\wedge A)\vee(P\wedge B)$$

$$F\ (P\vee A)\wedge(P\vee B)$$

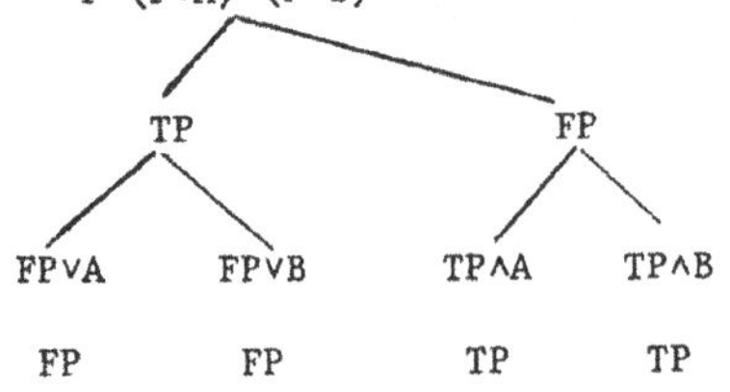

It is rather easy to see that any proof of this theorem without cuts must contain at least two more formulas (there must be a β rule application on the main branch, and then a β rule application on both new branches, and then deductions of TP and FP on each branch). For instance, no cut-free proof could be shorter than the following:

$$F\ [(P\wedge A)\vee(P\wedge B)]\supset[(P\vee A)\wedge(P\vee B)]$$

$$T\ (P\wedge A)\vee(P\wedge B)$$

$$F\ (P\vee A)\wedge(P\vee B)$$

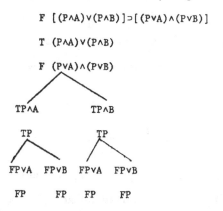

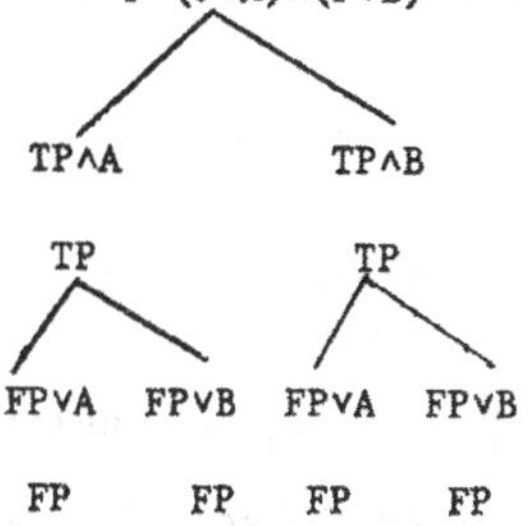

When we look at this proof, however, we notice two duplicate subtrees. After considering this example we might wonder whether cut elimination in propositional logic, and perhaps also even in first order logic, might be looked at as repetition-introduction. Not surprisingly (considering we construct cut-free proofs from proofs with cuts), the answer turns out to be yes, strictly so in propositional logic, in a more general sense in quantification theory. Thus the **Hauptsatz for first order logic** given below has the following meaning. We assume we are given a closed first order tableau that may have cuts. Let us mark all the cut formulas and their descendents and then look at the remaining unmarked formulas. If we give all such formulas distinct names or labels $L_1,\ldots,L_R$, we can consider a sequence of points labelled $L_{i_1},\ldots,L_{i_n}$ a subargument of $\mathcal{T}$ if it contains the nonmarked points on some branch of $\mathcal{T}$, starting at the origin. The idea is that we can transform the tableau $\mathcal{T}$ to a cut-free tableau $\mathcal{T}'$ in which every point has one of the labels $L_1, \ldots L_R$, in such a way that (1) two points labelled L_i are the same modulo a change in parameters; (2) if $L_{i_{j-1}}, L_{i_j}$ occurred in a subargument of $\mathcal{T}$, every point labelled L_{i_j} in $\mathcal{T}'$ has above it a point labelled $L_{i_{j-1}}$, if L_i came from L_j in $\mathcal{T}$, every point labelled L_i comes from a point labelled L_j in $\mathcal{T}'$; i.e., the subarguments of $\mathcal{T}$ are preserved in $\mathcal{T}'$ (modulo uniform changes of parameters) although they may be intertwined and some may have been eliminated. This means also that if we ignore repetitions of the same argument in calculating the size of the two tableaux <u>the cut-free tableau is actually smaller</u>, and smaller by at least the number of cut formulas and their descendents that occurred in the original tableau. Now for the details.

Let $\{X_i\}$ denote a finite sequence of points $X_1, \ldots, X_n$. A sequence $\{X_i'\}$ is said to be a <u>specialization</u> of the sequence $\{X_i\}$ if $\{X_i'\}$ can be obtained by simultaneously substituting certain parameters $b_1, \ldots, b_k$ for certain distinct parameters $a_1, \ldots, a_k$ throughout all the formulas of $\{X_i\}$.

Let S be a finite set of formulas. For briefness, a finite closed tableau $\mathcal{T}$ for S will be called an S-tableau.

Now let $\mathcal{T}$ be an S-tableau. Let $X_1, \ldots, X_n$ be a sequence of points in $\mathcal{T}$, which we denote also by $\{X_i\}$ or $\{X_i\}_{i=1}^{n}$. We say $\{X_i\}$ is <u>pure</u> if no member of the sequence is a cut formula or descendent of a cut formula in $\mathcal{T}$. We say a pure sequence $\{X_n\}$ is a <u>subargument</u> if (1) X_{i+1} occurs below X_i in $\mathcal{T}$ $(i = 1,\ldots,n-1)$ and (2) every formula X_i is either an element of S or is the result of the application of a rule to a formula X_j where $j < i$. (For instance if a subargument $\{X_i\}$ contains only one member of S, that member must be the first element of $\{X_i\}$ and all other elements are descendents of it.)

Now let us be given a list P of m (not necessarily different) formulas $P_1,\ldots,P_m$, a list L of m formula names or labels, $L_1,\ldots,L_m$, and a set A of k sequences $A_1,\ldots,A_k$, each of which is a finite sequence of formulas from P. We will say a tableau $\mathcal{T}$ is P-L-A labelled in the following situation: Every formula in $\mathcal{T}$ that is not a cut formula or the descendent of a cut formula and no other formula, has a label from L associated with it. Also each such point X has associated with it a subargument $\{X_i\}$ of which it is a member. If two points X and X' have the same label, then (1) their subarguments $\{X_i\}$ and $\{X_i'\}$ have the same number of elements, say n, (2) corresponding members of their subarguments have the same labels, and (3) if $L_{j_1},\ldots,L_{j_n}$ is the sequence of labels corresponding to the subargument points, then both $\{X_i\}$ and $\{X_i'\}$ are specializations of the sequence $P_{j_1},\ldots,P_{j_n}$, which is an element of A.

The *base* of a tableau $\mathcal{T}$ is defined to be the maximal subtree of $\mathcal{T}$ containing the origin but no cut formula. We say the base B of a P-L-A labelled tableau $\mathcal{T}$ occurs at the top of a P-L-A labelled tableau $\mathcal{T}'$ if a subtree of $\mathcal{T}'$ containing the origin has the same formulas and structure as B *and* is labelled in the same way.

We are now prepared to prove that we can transform a P-L-A labelled S-tableau with cuts to a cut-free P-L-A labelled S-tableau. We will use two lemmas to prove this and actually obtain a stronger result.

Let $\mathcal{T}$ be a P-L-A labelled S-tableau. We say a tableau $\mathcal{T}'$ is a *$\mathcal{T}$-reduct* if $\mathcal{T}'$ is a P-L-A labelled S-tableau such that every label associated with a point of $\mathcal{T}'$ is also associated with some point of $\mathcal{T}$.

Lemma 1. Given a P-L-A labelled S-tableau $\mathcal{T}$ of degree 0 and rank $\leq r$, one can construct a $\mathcal{T}$-reduct $\mathcal{T}'$ such that (1) $\mathcal{T}'$ is cut-free, (2) $\mathcal{T}'$ is of rank $\leq 2^r$, (3) the base B of $\mathcal{T}$ occurs at the top of $\mathcal{T}'$, (4) if a label is used only in B in $\mathcal{T}$, it is used only in B in $\mathcal{T}'$.

Proof. We use induction on r. If r is 0, $\mathcal{T}$ itself is cut-free and satisfies (1), (2), (3) (4) as $\mathcal{T}'$.

Let r be greater than 0 and assume the theorem true of all tableau of rank $< r$ meeting the assumption requirements. If $\mathcal{T}$ has no cuts we are again done, so let us consider the contrary case.

Let us denote by θ the origin branch of the tableau. Thus $\mathcal{T}$ must be of the form

(A)

$$\begin{array}{c} \theta \\ | \\ y_1 \\ \mathcal{T}_1 \end{array}$$

or

(B)

$$\begin{array}{ccc} & \theta & \\ / & & \backslash \\ y_1 & & y_2 \\ \mathcal{T}_1 & & \mathcal{T}_2 \end{array}$$

where the y_1 or y_1 and y_2 are the result of some rule application. We can consider a tableau of form (A) as a tableau for the set $S\cup\{y_1\}$, in which case it has rank $\le r-1$. We then apply the induction hypothesis to obtain a cut-free $\mathcal{T}$-reduct $\mathcal{T}'$ for $S\cup\{y_1\}$ of rank $\le 2^{r-1}$ where the base of $\mathcal{T}$, which must contain

$$\begin{array}{c} \theta \\ | \\ y_1, \end{array}$$

occurs at the top of $\mathcal{T}'$. We can thus look at $\mathcal{T}'$ also as a tableau for S of rank $\le 2^{r-1} + 1 \le 2^r$ satisfying the requirements of the theorem.

Now let us consider a tableau of form (B). We divide it into two tableaux:

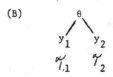

$$(*_1)\quad \begin{array}{c}\theta\\ y_1\\ \mathcal{T}_1\end{array} \qquad \text{and} \qquad (*_2)\quad \begin{array}{c}\theta\\ y_2\\ \mathcal{T}_2\end{array}$$

which we look at as tableaux for the sets $S\cup\{y_1\}$ and $S\cup\{y_2\}$ respectively. As such they are again of ranks $\leq r_1$, $\leq r_2$, where r_1 and r_2 are less than r. But we may not have labelled tableaux if y_1 and y_2 are cut formulas. But if this is so, then for $i = 1, 2,$ we take new labels and assign a distinct one of them to y_i and each of its descendents. We add copies of the formulas labelled to P, and for every formula thus newly labelled, we add to A, say, the list of its immediate ancestors, up to y_i, in reverse order. We thus obtain a new labelled tableau, which we can say is

$$P^{y_i}{}_{-L}{}^{y_i}{}_{-A}{}^{y_i}$$

labelled. We can now apply the induction hypothesis separately to the two labelled tableaux to obtain cut-free reducts:

$$\begin{array}{c}\theta\\ y_1\\ \mathcal{T}_1'\end{array} \qquad \text{and} \qquad \begin{array}{c}\theta\\ y_2\\ \mathcal{T}_2'\end{array}$$

of ranks $\leq 2^{r_1}$ and $\leq 2^{r_2}$ respectively and with the bases of the original tableaux at the tops. Now let us put these tableaux together to form:

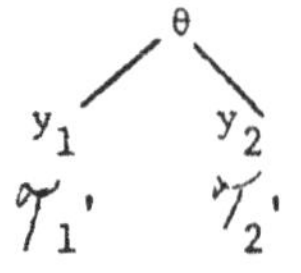

This tableau has rank $\leq \max\{2^{r_1} + 1, 2^{r_2} + 1\} \leq 2^r$ where $\mathcal{T}_1'$ and $\mathcal{T}_2'$ have no cuts. If y_1 and y_2 were not cut formulas, we now have a $\mathcal{T}$-reduct meeting all our requirements. If y_1 and y_2 were cut formulas we have a tableau with a single cut. Before seeing if we can eliminate this cut, however, we must verify if we can see the tableau above as a P-L-A labelled tableau or, more precisely, a $\mathcal{T}$-reduct. Recall we formed it from our P^{y_1}-L^{y_1}-A^{y_1} and P^{y_2}-L^{y_2}-A^{y_2} labelled reducts. Let us just eliminate the labels associated with y_1 and y_2 and their descendents. The question is whether any other formulas in (*) have the labels in L^{y_1} and L^{y_2} that are not in L. The answer is no because of the final clause in the induction hypothesis and the form of the subarguments we introduced in A^{y_1} and A^{y_2}. To consider this in more detail, let i be 1 or 2 and let X be a formula in

$$(**_i) \qquad \begin{array}{c} \theta \\ y_i \\ \mathcal{T}_i', \end{array}$$

that has a label in L^{y_i} not in L. Now X has a subargument $\{X_j\}$ in $(**_i)$ that must be a specialization of one of the sequences of A^{y_i} that is not in A. But the first element in any such sequence is y_i, so the first element of $\{X_j\}$ must also have the same label as y_i. The induction hypothesis, however, tells us that if a label is used only in the base B of $(*_i)$, it occurs only in B in $(**_i)$ and that B occurs at the top of $(**_i)$. Now y_i is in B in $(*_i)$ and is the only formula with its label, so the same must be true in $(**_i)$. Hence $\{X_j\}$ actually starts with y_i and X <u>is</u> a descendent of y_i.

Thus we have only to consider the case of a P-L-A labelled tableau (a $\mathcal{T}$-reduct)

$$(**) \qquad \begin{array}{ccc} & \theta & \\ \swarrow & & \searrow \\ y_1 & & y_2 \\ \mathcal{T}_1' & & \mathcal{T}_2' \end{array}$$

with one cut to y_1 and y_2 where y_1 and y_2 have orders $\leq 2^{r_1}$ and $\leq 2^{r_2}$ respectively, $r_1 < r$, $r_2 < r$.

In the case we are considering y_1 and y_2 are atomic formulas. We proceed to eliminate the cut formulas $y_1 = \overline{y_2}$ and y_2 as follows. We first form the tree

$$\begin{array}{c} \theta \\ \mathcal{T}_1'. \end{array}$$

This is clearly a P-L-A labelled tableau of order $\leq 2^{r_1} \leq 2^r$. If all its branches are closed, we are done. If not, the branches that are not closed come from branches in $(**_1)$ that needed y_1 to close (thus they contain $\overline{y_1} = y_2$.) At the end point of each such open branch we attach $\mathcal{T}_2'$:

$$\begin{array}{ccc} & \theta & \\ & \mathcal{T}_1' & \\ \swarrow & \downarrow & \searrow \\ \mathcal{T}_2' & \mathcal{T}_2' & \mathcal{T}_2' \end{array}$$

The new branches that result can be written $\theta, \theta_1, \theta_2$, where θ_1 is a path in $\mathcal{T}_1'$ containing y_2, and θ_2 is a path in $\mathcal{T}_2'$. In seeing that each such branch is closed we consider two cases. θ, θ_2 may itself contain some formula and its conjugate. Otherwise y_2 was needed in $(**_2)$ to close θ, y_2, θ_2, so θ, θ_2, contains $\bar{y}_2$. But θ, θ_1 contains y_2 and again the branch is closed. It is easy to see that the result is a cut-free P-L-A labelled tableau and, more specifically a $\mathcal{T}$-reduct. Now the ranks of some of the end points in the tree

$$\begin{array}{c} \theta \\ \mathcal{T}_1' \end{array}$$

have been changed from ≤ 0 to $\leq 2^{r_2}$. By the rank lemma 2, the resulting tree is of rank $\leq 2^{r_2} + 2^{r_1} \leq 2^{\max\{r_1,r_2\}} + 2^{\max\{r_1,r_2\}} = 2^{\max\{r_1,r_2\}+1} = 2^r$. So our resulting tableau is satisfactory in all ways and we have completed the proof of lemma 1.

Lemma 2. Given a P-L-A labelled S-tableau of degree $d > 0$ and rank $\leq r$, one can construct a $\mathcal{T}$-reduct $\mathcal{T}'$ such that (1) $\mathcal{T}'$ is of degree $< d$; (2) $\mathcal{T}'$ is of rank $\leq 2^r$; (3) the base B of $\mathcal{T}$ occurs at the top of $\mathcal{T}'$; (4) if a label is used only in B in $\mathcal{T}$ it is used only in B in $\mathcal{T}'$.

Proof. Again we use induction on r. In this case r cannot be 0. Let the r be > 0 and assume the theorem true for tableaux of rank $< r$ (and degree $\leq d$). The first part of this case is very similar to the first part of the induction case in the proof of lemma 1, i.e., through the forming of the P-L-A labelled S-tableau (and $\mathcal{T}$-reduct):

$$(**) \qquad \begin{array}{ccc} & \theta & \\ \swarrow & & \searrow \\ y_1 & & y_2 \\ \mathcal{T}_1' & & \mathcal{T}_2' \end{array}$$

where y_1 and y_2 are cut formulas and the tableau is of order $\leq 2^r$. In this case, however, $\mathcal{T}_1'$ and $\mathcal{T}_2'$ are not necessarily cut-free, but may contain no cuts of degree $\geq d$ (throughout the part of the proof of lemma 1 referred to we need only replace "cut-free" by "of degree $< d$"). Now if the cut to y_1 and y_2 is of degree $< d$ we are done, for (**) is a $\mathcal{T}$-reduct satisfying (1), (2), (3), (4). Let us assume otherwise. We have two cases depending on the kinds of cut formulas.

Case 1. We have a propositional cut, say y_1 is an α and y_2 is $\bar{\alpha}$.

If the β rule is never applied to $\bar{\alpha}$ in $\mathcal{T}_2'$, the tableau

$$\begin{array}{c} \theta \\ \mathcal{T}_2' \end{array}$$

satisfies the requirements of the theorem. Similarly, if the

α rule is never applied to α in $\mathcal{T}_1'$ the tableau

$$\begin{array}{c}\theta\\ \mathcal{T}_1'\end{array}$$

satisfies (1), (2), (3), (4). Now let us assume both α and $\bar{\alpha}$ are used somewhere in $\mathcal{T}_1'$ and $\mathcal{T}_2'$ respectively. Now we construct two trees, $\mathcal{T}^{\bar{\alpha}_1}$ and $\mathcal{T}^{\bar{\alpha}_2}$ from $\mathcal{T}_2'$ as follows. For $\mathcal{T}^{\bar{\alpha}_1}$ ($\mathcal{T}^{\bar{\alpha}_2}$) we first eliminate any formula in $\mathcal{T}_2'$ that occurs below an $\bar{\alpha}_2$ ($\bar{\alpha}_1$) that is a descendent of $\bar{\alpha}$ and then we eliminate all the $\bar{\alpha}_1$ and $\bar{\alpha}_2$ that come from $\bar{\alpha}$. We call this intermediate result $\mathcal{T}_{21}'$ ($\mathcal{T}_{22}'$).

Then if $a_1,\dots,a_n$ $(b_1,\dots,b_m)$ are the parameters introduced by the δ rule in $\mathcal{T}_2'$, we choose n (m) parameters $c_1,\dots,c_n$ $(d_1,\dots,d_m)$ new to the tableau (**) and replace $a_1,\dots,a_n$ $(b_1,\dots,b_m)$ by $c_1,\dots,c_n$ $(d_1,\dots,d_m)$ throughout $\mathcal{T}_{21}$ ($\mathcal{T}_{22}$) to form $\mathcal{T}^{\bar{\alpha}_1}$ ($\mathcal{T}^{\bar{\alpha}_2}$). We keep the same labels on all formulas. Thus $\mathcal{T}^{\bar{\alpha}_i}$ is the part of $\mathcal{T}_2'$ that explores the $\bar{\alpha}_i$ possibilities, although $\bar{\alpha}_i$ has been removed. We now form a new tableau $\mathcal{T}'$ by placing θ at the top, then $\mathcal{T}_1'$, and then, wherever the α rule is used in

$$\begin{array}{c}\theta\\ \alpha\\ \mathcal{T}_1'\end{array}$$

to obtain an α_i in $\mathcal{T}_1'$, we make a cut

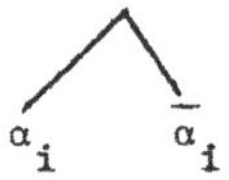

and place $\gamma^{\overline{\alpha}_i}$ under $\overline{\alpha}_i$. Thus we might have, for instance:

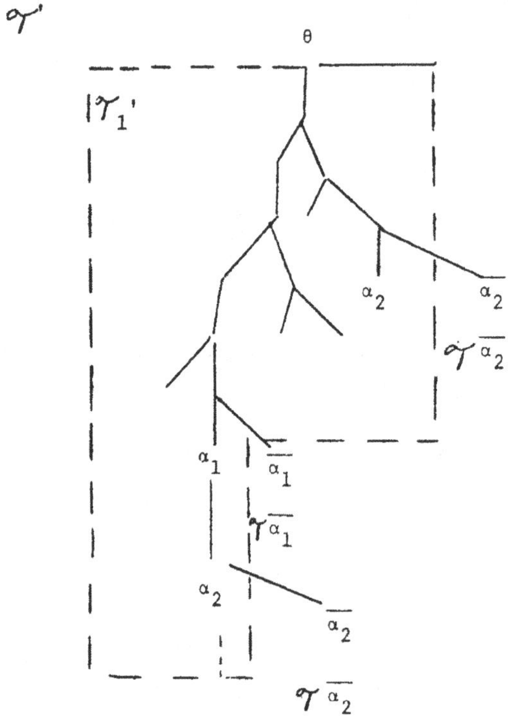

We must first see that we still have a γ-reduct. Clearly, subarguments are maintained because any formula that occurs in γ' still has above it all those formulas it originally did except for cut formulas and their descendents, which are never labelled and did not occur in subarguments. And the change of parameters that may have occurred still leaves the subargument a specialization of the same member of A. The tree is still a tableau - i.e., every point in it is either a member of S or the result of a rule. For every point in $\mathcal{T}_1'$ is still derived as before with the exceptance of the α_i, which are now obtained with a cut rule instead of the α rule. And every point in $\mathcal{T}^{\bar{\alpha}_i}$ either come from some point in θ, or from $\bar{\alpha}_i$, or from a point within itself; and α_i occurs above every occurrence of $\mathcal{T}^{\bar{\alpha}_i}$. We have made the δ rule applications valid in their new position by our change of parameters. Thus all points are validly derived by rules. We must now see that the tableau is closed. The set of points on any branch in it is of one of two forms: a) the set of points from some branch in

$$\begin{array}{c}\theta\\ \alpha\\ \mathcal{T}_1'\end{array}$$

except for the point α; b) the set of points from some branch in

$$\begin{array}{c}\theta\\ \bar{\alpha}\\ \mathcal{T}_2'\end{array}$$

except for the $\bar{\alpha}$, plus some points from some branch in $\mathcal{T}_1'$. We see that deleted points are always nonatomic, and thus could not have been used for closure. It remains to see that the rank of the tableau is correct, but this follows from rank lemma 2.

Case 2. We have a first order cut, say y_1 is a γ and y_2 is $\bar{\gamma}$. This case is very similar to the first case. We will go into detail, however, about the tableau constructed when both γ and $\bar{\gamma}$ are used somewhere in $\mathcal{T}_1'$ and $\mathcal{T}_2'$ respectively.

So let e be any parameter. We can construct a tree $\mathcal{T}^e$ from $\mathcal{T}_2'$ as follows. First, for every parameter d introduced into $\mathcal{T}_2'$ by the δ rule applied to $\bar{\gamma}$, we replace d by e throughout $\mathcal{T}_2'$. We then eliminate all occurrences of $\bar{\gamma}(e)$ in the result. Let us call this intermediate result $\mathcal{T}_2'e$. Then if $a_1,\ldots,a_n$ are the other parameters introduced by the δ rule in $\mathcal{T}_2'$, we choose n parameters $b_1,\ldots,b_n$ different from e and new to the tableau (**) and replace $a_1,\ldots,a_n$ by $b_1,\ldots,b_n$ throughout $\mathcal{T}_2'e$ to form $\mathcal{T}^e$. Thus $\mathcal{T}^e$ explores the possibilities of $\bar{\gamma}(e)$ in the situations where $\mathcal{T}_2'$ explored $\bar{\gamma}(d)$ for various arbitrary parameters d. We now form a new tableau $\mathcal{T}'$ by placing θ at the top, then $\mathcal{T}_1'$ and then, wherever the γ rule is used in

$$\begin{array}{c}\theta\\ \gamma\\ \mathcal{T}_1'\end{array}$$

to obtain a $\gamma(e)$ in $\mathcal{T}_1'$, we make a cut

$$\gamma(e) \qquad \overline{\gamma(e)}$$

and place $\mathcal{T}^e$ under $\bar{\gamma}(e)$. We see as in the previous case that the tree formed in a $\mathcal{T}$-reduct satisfying all the requirements of the theorem.

This completes the proof of lemma 2.

Let $2_d(r)$ stand for

$$d \left\{ 2^{2^{\cdot^{\cdot^{\cdot^{2^{r}}}}}} \right.$$

Our complete result upon combining lemmas 1 and 2 is:

Theorem. Given a P-L-A labelled S-tableau $\mathcal{T}$ of degree d and rank $\leq r$, one can construct a cut-free $\mathcal{T}$-reduct of rank $\leq 2_{d+1}(r)$ (where the base B of the first occurs at the top of the second and any label used only in B in the first is used only in B in the second.)

Let $\mathcal{P}(\mathcal{T})$ be the number of points in $\mathcal{T}$. Let $\mathcal{L}(\mathcal{T})$, the label size of $\mathcal{T}$ (with respect to a particular labelling), be the number of points in $\mathcal{T}$ not counting more than once points with the same label (i.e., $\mathcal{L}(\mathcal{T})$ is the number of labels assigned to points in $\mathcal{T}$ plus the number of cut formulas and their descendents in $\mathcal{T}$). Let $\mathcal{C}(\mathcal{T})$ be the number of cut formulas and their descendents in $\mathcal{T}$. We have:

Theorem. Let a S-tableau $\mathcal{T}$ (with cuts) be given. One can find a labelling for $\mathcal{T}$, say by sets P, L, A where $\mathcal{P}(\mathcal{T}) = \mathcal{L}(\mathcal{T})$. Then one can find a cut-free $\mathcal{T}$-reduct $\mathcal{T}'$ where

$\mathcal{L}(\mathcal{T}') \leq \mathcal{L}(\mathcal{T}) - \mathcal{C}(\mathcal{T})$.

Proof. Let $\mathcal{T}$ be an S-tableau, perhaps with cuts. First number the points in $\mathcal{T}$ that are not cut formulas or their descendents. Select a distinct label L_i for each such point X_i in $\mathcal{T}$. Let L consist of the sequence of labels thus formed. Let P consist of the sequence of points X_i. For each point X_i, let A_i be the sequence of all the points from the origin to X_i except for cut formulas and their descendents. Let A be the set of A_i. Then $\mathcal{T}$ is a P-L-A labelled S-tableau in which $\mathcal{P}(\mathcal{T}) = \mathcal{L}(\mathcal{T})$. By theorem 1 we can now find a cut-free $\mathcal{T}$-reduct $\mathcal{T}'$. Since at most the same labels are used and only labelled formulas occur in $\mathcal{T}'$, clearly $\mathcal{L}(\mathcal{T}') \leq \mathcal{L}(\mathcal{T}) - \mathcal{C}(\mathcal{T})$.

It is this theorem that expresses the intuitive idea that the cut-free tableau we obtain from a tableau with cuts may never be larger, if repetitions are considered. In fact, it <u>must</u> be smaller and by at least the number of cut formulas and their descendents that occurred in the tableau with cuts.

We note that this constructive proof of cut-elimination provides a constructive proof of the syntactic consistency of first order logic as follows.

<u>Theorem 51</u>. There is no formula A such that both A and $\neg A$ are provable in first order logic.

<u>Proof</u>. Assume the contrary. Then there is a closed tableau

$$\begin{array}{c} F\ A \\ \mathcal{T}_1 \end{array}$$

for A and a closed tableau

$$\begin{array}{c} F\neg A \\ \mathcal{T}_2 \end{array}$$

for $\neg A$. Now let B be an arbitrary formula whose parameters are not in any of the tableaux mentioned. We have the following proof of B:

$$\begin{array}{ccc} & F\,B & \\ \swarrow & & \searrow \\ T\neg A & & F\neg A \\ F\,A & & \mathcal{T}_2 \\ \mathcal{T}_1 & & \end{array}$$

By our cut elimination theorem we can obtain from this proof cut-free proof of B. But it is clear we cannot prove any atomic B without cuts, for no rules can be used. We have thus reached a contradiction and the theorem must be true.

One should see whether one can find a counter-example to this theorem as it stands in second order logic. The question of whether the theorem would hold if terms were allowed to be substituted for parameters is closely related to the question of whether there is a constructive Hauptsatz at all.

APPENDIX 2 - A Translation Procedure Between Tableau Systems and Schütte Systems with Positive and Negative Parts

In this appendix we will show how to translate a proof in a tableau system such as we have given into a proof in a system such as Schütte uses in Beweistheorie and vice-versa. For simplicity, we will do this in terms of systems for pure first order logic. The modifications needed to handle the systems we have discussed above are not difficult to work out.

Schütte's systems are based on the concept of the positive and negative parts of a formula. These are subformulas of a formula. More precisely, a positive (negative) part of a formula is a component of the formula from whose truth (falsity, respectfully), the truth of the whole formula follows by the truth definition of propositional logic. We may define the positive and negative parts of a formula A inductively as follows:

1) A is a positive part of A.

2) If $\neg$ B is a positive part of A, then B is a negative part of A.

3) If $\neg$ B is a negative part of A, then B is a positive part of A.

4) If (B $\vee$ C) is a positive part of A, then both B and C are positive parts of A.

5) If (B $\wedge$ C) is a negative part of A, then both B and C are negative parts of A.

6) If (B $\supset$ C) is a positive part of A, then B is a negative part of A and C is a positive part of A.

When we write A[B+] or C[B-] we denote a formula with respect to one occurrence of a positive or negative part B. In such a context, then A[D+] or C[D-] denotes the result of replacing the positive or negative part B by D in the formula A[B+] or C[B-]. Similarly A[D+,E+], B[D-, E-] or C[D+, E-] denotes a formula with respect to one occurrence of a positive or negative part D and one occurrence of a positive or negative part E. If A[B] is a formula in which a positive or negative part B occurs, we denote by A[ ] the result of "deleting" B from A[B] and then eliminating unnecessary parentheses and connectives. This procedure can be described inductively as follows:

a) Eliminate B from A[B].

b) When a symbol of A[B] is deleted that directly follows a $\neg$, also delete the $\neg$.

c) If A[B] has a component $(C \vee D)$, $(D \vee C)$, $(C \wedge D)$, $(D \wedge C)$, $(C \supset D)$, or $(D \supset C)$, and C is deleted, replace the component by D (i.e., delete the connective and the parenthesis also).

One can easily see that the result of deleting a positive or negative part of a formula is either a formula, or the empty string of symbols, which we denote here by ϕ.

We say a tableau is a complete α tableau for a formula if only α rules are applied in the tableau and if an α formula occurs in the tableau then both its successors occur in the tableau. It is clear than any complete α tableau consists of a single branch.

We have:

Lemma 1. B is a positive (negative) part of A iff F B (T B, respectively) occurs in every complete α tableau for F A.

The proof follows the inductive definition of positive and negative parts.

We can now describe Schütte's system for first order logic, which he calls Π, as follows:

The axioms of Π are all the formulas of the form $A[B+,B-]$, where B is an atomic formula.

The rules are the following (here the formula or formulas to the left of the arrow are the premisses of the rule while the formula to the right of the arrow is the conclusion of the rule):

S1a. $A[B+], A[C+] \to A[(B \wedge C)+]$,

S1b. $A[B-], A[C-] \to A[(B \vee C)-]$,

S1c. $A[(\neg B)-], A[C-] \to A[(B \supset C)-]$.[1]

S2a. $A[B(a)+] \to A[(\forall x)B(x)+]$,

S2b. $A[B(a)-] \to A[(\exists x)B(x)-]$,

where a does not occur in the conclusion.

S3a. $A[(\exists x)B(x)+] \vee B(a) \to A[(\exists x)B(x)+]$,

S3b. $A[(\forall x)B(x)-] \vee \neg B(a) \to A[(\forall x)B(x)-]$.

S. $A[C+], B[C-] \to A[\ ] \vee B[\ ]$.

We will see later that the axioms correspond to the closure sets and the α rules in tableaux, the S1 rules to the β rules, the S2 rules to the δ rules, the S3 rules to the γ rules, and the S rule to the cut rule.

Here in the S rule either $A[\ ]$ or $B[\ ]$ or both may be ϕ. To obtain a similar tableau system, we consider the system Π'

[1] Schütte's systems do not actually use the implication sign as a primitive.

that is like the system described by Smullyan in [1], (i.e., the rules are the $\alpha, \beta, \gamma, \delta$, and cut rules, the closure sets {T A, F A}, where A is atomic) except that a tableau is allowed to begin with a cut. It is easy to see that a finite or empty set S of formulas has a closed tableau in Π' iff it does in Smullyan's system - let us call it Σ - in which cuts may not begin tableaux. For if a formula from S is used in a Π' tableau, it may be transferred to the top to yield an S tableau. Otherwise, an atomic formula A may be found without parameters in the tableau, and F A may be placed at the top of the tableau to give a proof of A in Σ. But this is impossible by the cut elimination theorem for Σ and a consideration of the tableau rules. In this chapter we will consider only tableaux that are proofs of formulas. We will say a tableau for an empty set of formulas is a proof of ϕ. We note that the S rule might also conceivably lead us to a proof of ϕ in Π.

Our first goal is to show that anything provable in the Schütte system Π is provable in the Smullyan system Π'. In fact, if we call a tableau <u>nonrepetitive</u> when no rule is allowed to be applied to the same formula to yield the same conclusion(s) more than once on any given branch, we will show everything provable in Π has a non-repetitive proof in Π'. To do this we will use a few additional concepts and lemmas.

Let us denote by $\mathcal{T}_{A[C+]}$ or $\mathcal{T}_{A[C-]}$ a nonrepetitive tableau for F A[C+] or F A[C-] in which (1) only α rules are applied, (2) an α rule is applied (twice) to every formula except F C, and

and T C respectfully, and (3) the α rule is not applied to F C or T C (a "complete" α tableau except for F C or T C). Then we denote by $\mathcal{T}_{A[D+]}$ or $\mathcal{T}_{A[D-]}$ the nonrepetitive tableau which is the result of replacing C by D in the initial formula and in all the formulas in the tableau that have an "image" of C. We denote by $\mathcal{T}_{A[\ ]}$ the result of deleting C in the initial formula and in all the formulas of the tableau containing an "image" of C and then deleting repetitious rule applications. Here some signed formulas or even the whole tableau may disappear altogether.

It is easy to see that under such assumptions:

Lemma 2. Let D be an arbitrary formula. If S_1 is the set of non-α formulas in $\mathcal{T}_{A[C+]}$ and S_2 is the set of non-α formulas in $\mathcal{T}_{A[D+]}$, then $S_1 - \{F\ C\} = S_2 - \{F\ D\}$. If S_1 is the set of non-α formulas in $\mathcal{T}_{A[C-]}$ and S_2 is the set of non-α formulas in $\mathcal{T}_{A[D]}$, then $S_1 - \{T\ C\} = S_2 - \{T\ D\}$.

Lemma 3. If S_1 is the set of non-α formulas in $\mathcal{T}_{A[C+]}$ and S_2 is the set of non-α formulas in $\mathcal{T}_{A[\ ]}$, then $S_1 - \{F\ C\} = S_2$. If S_1 is the set of non-α formulas in $\mathcal{T}_{A[C-]}$, and S_2 is the set of non-α formulas in $\mathcal{T}_{A[\ ]}$, then $S_1 - \{T\ C\} = S_2$.

Lemma 4. If $\mathcal{T}$ is a nonrepetitive tableau for F A and A is provable $\mathcal{T}$ can be extended to a nonrepetitive proof of A.

Now we can prove:

<u>Theorem 52</u>. If A is provable in Π, A has a nonrepetitive proof in Π'.[2]

Proof. We show that the axioms of Π are provable in Π', and that if the premisses of a rule of Π are provable in Π', so is the conclusion.

So first let A be an axiom of Π, i.e., let it be a formula B[C+,C-], where C is an atomic formula. By lemma 1, both T C and F C occur in the complete α tableau for F A, so that tableau is a proof of A in Π'.

Now let A be the consequence of an S1a rule, where the first premiss is B[C+], the second premiss is B[D+], and the conclusion A, is B[(C ∧ D)+]. We assume B[C+] and B[D+] have nonrepetitive proofs in Π'. Let $\mathcal{T}_{B[(C \wedge D)+]}$ be a non-repetitive complete α tableau for F B[(C ∧ D)+], and hence a tableau satisfying the assumptions of lemmas 2 - 4. By lemma 4 we can find closed nonrepetitive tableaux

(i) $\begin{array}{c}\mathcal{T}_{B[C+]}\\ \mathcal{T}_1\end{array}$ and (ii) $\begin{array}{c}\mathcal{T}_{B[D+]}\\ \mathcal{T}_2\end{array}$

Let us consider the tree

(iii)

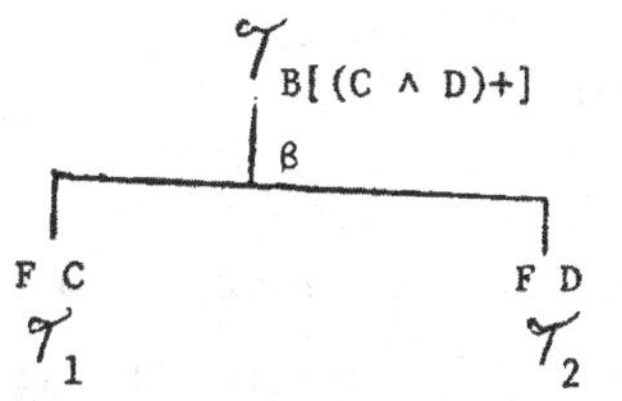

---

[2]Here A may be ϕ.

Since we assume the tableaux (i) and (ii) are nonrepetitive, all the formulas in $\mathcal{T}_1$ are derived from non-α formulas of $\mathcal{T}_{B[C+]}$ or from F C (all α formulas other than F C have the rule applied to them in $\mathcal{T}_{B[C+]}$ to obtain both possible results) and all the formulas in $\mathcal{T}_2$ are derived from non-α formulas of $\mathcal{T}_{B[D+]}$ or from F D. We apply lemma 2 to see that (iii) is a nonrepetitive tableau. It is closed because the atomic formulas in any branch of it are the same as the atomic formulas of some branch in (i) or (ii). Thus (iii) is a nonrepetitive proof of $B[(C \wedge D)+]$ in Π'.

The case when A is the consequence of an S1b rule is similar to the case of an S1a rule.

Let A be the conclusion of an S1c rule, where the first premiss is $B[(\neg C)-]$, the second premiss is $B[D-]$, and A is $B[(C \supset D)-]$. We assume $B[(\neg C)-]$ and $B[D]$ have nonrepetitive proofs in Π'. Let $\mathcal{T}_{B[(C \supset D)-]}$ be a tableau for $F\ B[(C \supset D)-]$ satisfying the conditions of lemmas 2 - 4. By lemma 4 we can find closed nonrepetitive tableaux

(i) $\begin{array}{c}\mathcal{T}_{B[(\neg C)-]}\\ \mathcal{T}_1\end{array}$ and (ii) $\begin{array}{c}\mathcal{T}_{B[D-]}\\ \mathcal{T}_2\end{array}$

Consider the tree

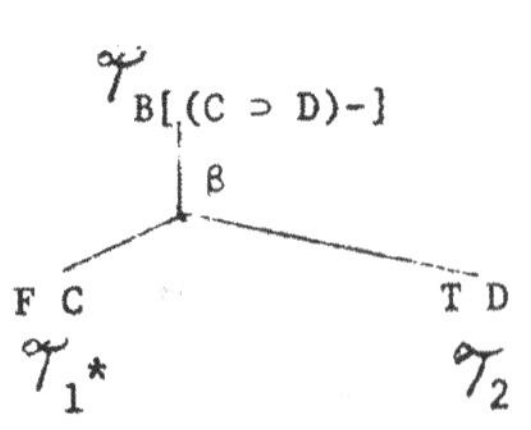

where $\mathcal{T}_1{}^*$ is the result of eliminating occurrences of F C in $\mathcal{T}_1$. We can see as in the case of rule S1a that this tableau is a nonrepetitive proof of $B[(C \supset D)-]$ in Π'.

Let A be the conclusion of an S2a rule, where the premiss in B[C(a)+] and A is B[(∀x)A(x)+], and a does not occur in A. We assume B[C(a)+] has a nonrepetitive proof in Π'. Let $\mathcal{T}_{B[(\forall x)A(x)+]}$ be a nonrepetitive tableau for F B[(∀x)A(x)+] satisfying the assumptions of lemmas 2 - 4. By lemma 4 we can find a closed nonrepetitive tableau

$$\begin{array}{c} \mathcal{T}_{B[C(a)+]} \\ \mathcal{T}_1 \end{array}$$

Clearly the parameter a cannot be introduced in $\mathcal{T}_1$ by a δ rule. It is then easy to see as in the case of rule S1a that

$$\begin{array}{c} \mathcal{T}_{B[(\forall x)C(x)+]} \\ \Big| \;\delta \\ F\ C(a) \\ \mathcal{T}_1 \end{array}$$

is a nonrepetitive proof of B[(∀x)C(x)+] in Π'.

The case when A is the conclusion of an S2b rule is similar.

Let A be the conclusion of an S3b rule, where the premiss is B[(∀x)C(x)-] ∨ ¬C(a) and A is B[(∀x)C(x)-]. We assume the premiss has a nonrepetitive proof in Π'. It is easy to see there is a nonrepetitive complete α tableau for F B[(∀x)C(x)-] ∨ C(a) of the form

$$\begin{array}{c} F\ B[(\forall x)C(x)-] \vee \neg C(a) \\ \mathcal{T}_{B[(\forall x)C(x)-]} \\ F \neg C(a) \\ T\ C(a) \\ \mathcal{T}_1 \end{array}$$

Then by lemma 4 there is a nonrepetitive proof of $B[(\forall x)C(x)-] \vee \neg C(a)$ of the form

$$
\begin{array}{c}
F\ B[(\forall x)C(x)-] \vee \neg C(a) \\
\mathcal{T}_{B[(\forall x)C(x)-]} \\
F \neg C(a) \\
T\ C(a) \\
\mathcal{T}_1 \\
\mathcal{T}_2
\end{array}
$$

Then it is easy to see that

$$
\begin{array}{l}
\mathcal{T}_{B[(\forall x)C(x)-]} \\
\quad \Big| \ \gamma \\
T\ C(a) \\
\mathcal{T}_1 \\
\mathcal{T}_2
\end{array}
$$

is a nonrepetitive proof of $B[(\forall x)C(x)-]$.

The case when A is the conclusion of an S3a rule is similar.

Let A be the conclusion of a cut rule, where the first premiss is $B[D+]$, the second premiss is $C[D-]$, and A is $B[\ ] \vee C[\ ]$. We assume that $B[D+]$ and $C[D-]$ have nonrepetitive proofs in Π'. Then by lemma 4 $F\ B[D+]$ and $F\ C[D-]$ have closed nonrepetitive tableaux

$$
\begin{array}{c}
\mathcal{T}_{B[D+]} \\
\mathcal{T}_1
\end{array}
\qquad \text{and} \qquad
\begin{array}{c}
\mathcal{T}_{C[D-]} \\
\mathcal{T}_2
\end{array}
$$

where $\mathcal{T}_{B[D+]}$ and $\mathcal{T}_{C[D-]}$ are nonrepetitive tableaux for $F\ B[D+]$ and $F\ C[D-]$ satisfying the conditions of lemmas 2 - 4.

Then we can see as in the case of rule S1a (this time with the use of lemma 3) that

$$
\begin{array}{c}
F\ B[\ ] \vee C[\ ] \\
\mathcal{T}_{B[\ ]} \\
\mathcal{T}_{C[\ ]} \\
\text{CUT} \\
F\ D \qquad\qquad T\ D \\
\mathcal{T}_1 \qquad\qquad \mathcal{T}_2
\end{array}
$$

is a nonrepetitive proof of B[ ] ∨ C[ ] in Π'.

<u>We note that proofs without cuts in Π go into proofs without cuts in Π'.</u>

We next wish to show that anything provable in Π' is provable in Π. To do this easily we will need three more lemmas. As the proofs of these lemmas are given in Schütte's <u>Beweistheorie</u> (they are rather long and tedious), we do not repeat them here, but refer the reader to that text.

<u>Lemma 5</u>. The formula A ∨ (B ∨ C) is provable in Π with (without) cuts iff the formula (A ∨ B) ∨ C is provable in Π with (without) cuts.

To prove this, see part II, proposition 36 and §6.5 and part III, §9.1 of <u>Beweistheorie</u>.

<u>Lemma 6</u>. If B[C+,C+] is provable in Π with (without) cuts, so is B[C+,]. If B[C−,C−] is provable in Π with (without) cuts, so is B[C−,].

To prove this see part III, §9.1 of Beweistheorie.

Lemma 7. If A is provable in Π with (without) cuts, so is $A \vee B$.

For the proof of this see part III, §9.1 of Beweistheorie.

Theorem 53. If A is provable in Π', it is provable in Π.

Proof. Let $\mathcal{T}$ be a closed tableau for F A in Π'. If X is a point in $\mathcal{T}$, we denote by $X_1, \ldots, X_n$ the formulas on the path P_X in $\mathcal{T}$. If Y is T B (F B), we use Y^0 to denote $\neg B$ (B, respectively). We will prove the theorem by showing that for every point X in $\mathcal{T}$, $X_1^0 \vee \ldots \vee X_n^0$ (cf. lemma 5) is provable in Π'. We do this first for the end points of $\mathcal{T}$, and then show that if the assertion holds for the successors of a given point it holds for the point.

Let X be an end point of a branch θ in $\mathcal{T}$. Since θ is closed it must contain T B and F B for some atomic formula B. Then in $X_1^0 \vee \ldots \vee X_n^0$ some X_i is $\neg B$ and some X_j is B. Clearly the formula B occurs as both a positive and negative part of $X_1^0 \vee \ldots \vee X_n^0$, and the formula is an axiom of Π.

Now let X be a formula such that its successor is derived by an α rule in $\mathcal{T}$. Thus we wish to show

(*) $$X_1^0 \vee \ldots \vee X_n^0$$

is provable in Π, where some X_i is an α, and we know

(**) $$X_1^0 \vee \ldots \vee X_n^0 \vee \alpha_k^0$$

is provable in Π, where k is 1 or 2. We note that the following cases may occur:

α	α^0	α_1^0	α_2^0
T $B \wedge C$	$\neg(B \wedge C)$	$\neg B$	$\neg C$
F $B \vee C$	$B \vee C$	B	C
F $B \supset C$	$B \supset C$	$\neg B$	C
T $\neg B$	$\neg\neg B$	B	B
F $\neg B$	$\neg B$	$\neg B$	$\neg B$

But in each case B (or C) occurs both in α^0 and α_1^0 (or α_2^0) as a part of the same kind. Thus by lemma 6, the derivability of $X_1^0 \vee \ldots \vee X_n^0$ follows from the derivability of (**).

Next let X be a formula such that its successors are derived by a β rule in $\mathcal{T}$. Thus we wish to show

(1) $$X_1^0 \vee \ldots \vee X_n^0$$

is provable in Π, where some X_i is a β, and we know both

(2) $$X_1^0 \vee \ldots \vee X_n^0 \vee \beta_1^0$$

and

(3) $$X_1^0 \vee \ldots \vee X_n^0 \vee \beta_2^0$$

are provable in Π. Again we note the possibilities:

β	β^0	β_1^0	β_2^0
F $B \wedge C$	$B \wedge C$	B	C
T $B \vee C$	$\neg(B \vee C)$	$\neg B$	$\neg C$
T $B \supset C$	$\neg(B \supset C)$	B	$\neg C$
T $\neg B$	$\neg\neg B$	B	B
F $\neg B$	$\neg B$	$\neg B$	$\neg B$

We first show how to derive

(4) $$X_1^0 \vee \ldots \vee X_n^0 \vee \beta^0$$

from (2) and (3). In the first two cases formula (4) follows from formulas (2) and (3) by an application of rule S1a or S1b. In the third case we first see that

$$(5) \qquad X_1^0 \vee \ldots \vee X_n^0 \vee \neg\neg B$$

follows from formula (2) by lemmas 7 and 6. Then formula (4) follows from formula (3) and formula (5) by an application of rule S1c. In the fourth case formula (4) follows from formula (2) or (3) by lemmas 6 and 7, and in the fifth case formulas (2), (3), and (4) are all identical. Finally, in every case, formula (1) follows from formula (4) by lemma 6.

Next let the successor of X be derived by a γ rule in $\mathcal{T}$. Thus we know

$$X_1^0 \vee \ldots \vee X_n^0 \vee \gamma(a)^0$$

is derivable in Π and wish to see that

$$X_1^0 \vee \ldots \vee X_n^0$$

is derivable in Π. But the latter formula follows from the former by rule S3a (if γ if $F(\exists x)C(x)$) or rule S3b (if γ is $T(\forall x)C(x)$).

Now let the successor of X be derived by a δ rule in $\mathcal{T}$. Thus we know

$$X_1^0 \vee \ldots \vee X_n^0 \vee \delta(a)^0$$

is derivable in Π, where a does not occur in $X_1^0 \vee \ldots \vee X_n^0$. By rule S2a (if δ is $F(\forall x)C(x)$) or rule S2b (if δ is $T(\exists x)C(x)$)

$$X_1^0 \vee \ldots \vee X_n^0 \vee \delta^0$$

is derivable in Π. By lemma 6,

$$X_1^0 \vee \ldots \vee X_n^0$$

is derivable in Π.

Finally let the successors of X be obtained by a cut rule. Thus we know

$$X_1^0 \vee \ldots \vee X_n^0 \vee C$$

and

$$X_1^0 \vee \ldots \vee X_n^0 \vee \neg C$$

are derivable in Π. By the cut rule in Π

$$X_1^0 \vee \ldots \vee X_n^0 \vee X_1^0 \vee \ldots \vee X_n^0$$

is derivable in Π. By lemma 6,

$$X_1^0 \vee \ldots \vee X_n^0$$

is derivable in Π.

We note that proofs without cuts in Π' go into proofs without cuts in Π. Thus, since proofs without cuts in Π went into proofs without cuts in Π', the cut elimination theorems in our different systems are equivalent for proofs of formulas that are not empty. This cut elimination in Π yields consistency as it does in Π'.

APPENDIX 3 - Applications of Gentzen's (Second) Consistency Proof to Intuitionistic Number Theory and Analysis (Scarpellini)

In this appendix we will describe briefly some of the ideas that have gone into Scarpellini's application of Gentzen's methods to obtain results concerning certain formalizations of intuitionistic mathematics.

Given any tableau system for classical mathematics (such as the ones that have been considered in this work), one can obtain a corresponding intuitionistic system by making the following restriction on proofs: given a branch in a proof, only the lowest F-prefixed formula can be used to extend the branch (while any T-prefixed formula on the branch can always be so used).

Scarpellini first noted that the operations involved in Gentzen's second consistency proof (given here in §2 of Chapter I) could be modified so as to be applicable to intuitionistic proofs. Here it was necessary, incidentally, for him to use Gentzen's original version, which included an operation for eliminating conjugate α and β rules similar to our operation 3 for eliminating conjugate quantificational rules.

We will first consider in detail one example illustrating the refinements that must be made to Gentzen's methods to obtain the desired results for intuitionistic number theory. (Here and throughout this appendix we will assume for convenience that no uses are made of the replacement rule in the formal proofs - the changes that must be made to take them into consideration are trivial but a nuisance to describe.) Now the operation that Gentzen would make to the proof I

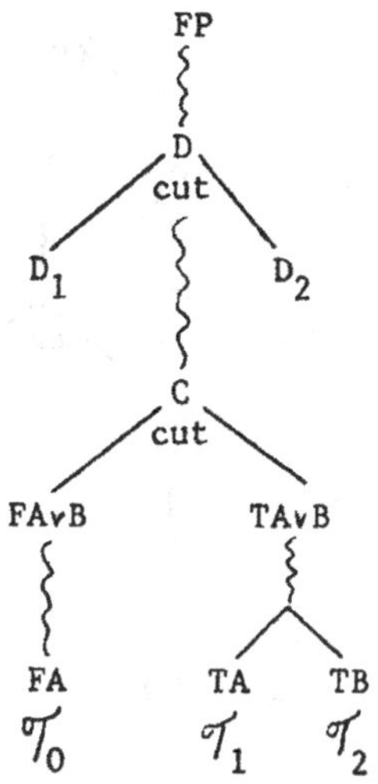

cut with depth change

I

(note here that assuming eliminations of all conjugate α - β's and forgetting about replacements has simplified the structure of what we have to work with; in particular, the only rule applied in the beginning part is the cut rule) would transform it to:

II

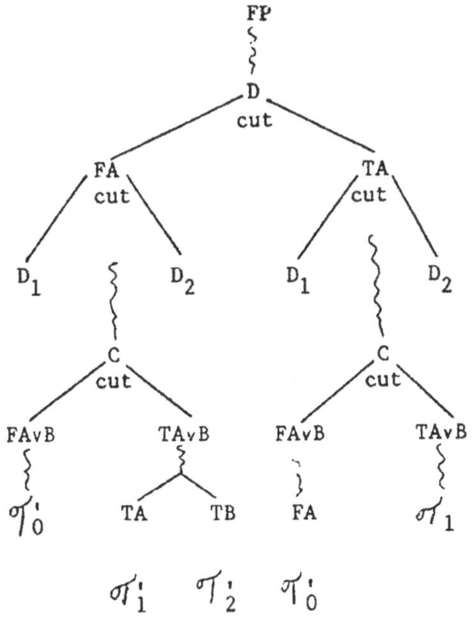

Here two different kinds of situations might arise to prevent the transformed proof II from being intuitionistic, given that the original proof I was. In the first place there may be an F-prefixed formula above the new FA that has to be used in the left-most π'_1. Thus the new position of FA would prevent this intuitionistically. The solution to this difficulty is to note that the FA∨B under FA is never used, so that the whole cut involving it can be eliminated, which gets rid of the troublesome π'_1 simultaneously:

III

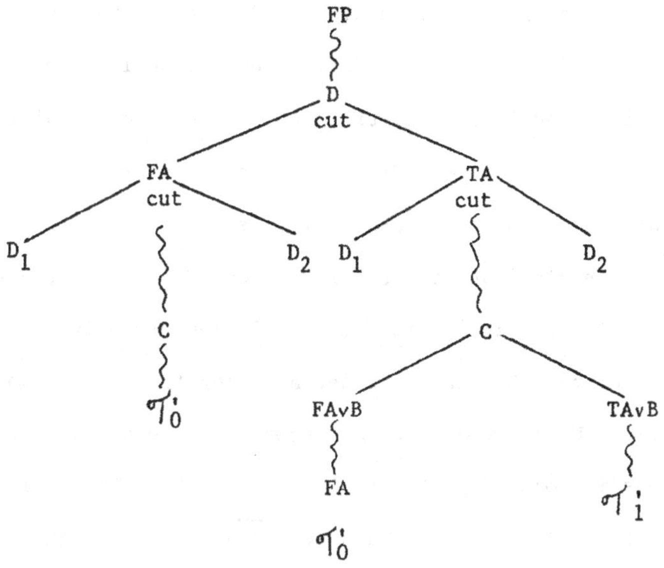

The only problem that can still remain is that there might be some F-prefixed formulas between the new FA and the places it is used in $\mathcal{T}'_0$. But if we look at the original tableau, we see that these formulas are cut formulas which could never have been used anywhere in $\mathcal{T}'_0$ (because of the old FA) or on the path above it (because that path is part of the beginning part). They can then just be completely eliminated from our proof III, giving us a proof IV that is an intuitionistic proof once again (and one with a lower ordinal than that of proof I, under the usual ordinal assignment).

Now exactly the same considerations apply for every quantifier and connective, thus giving the extension of Gentzen's method to intuitionistic proofs.

For our present purposes, we also need to note that an additional operation used by Gentzen, let us call it operation 1', can be applied to both classical and intuitionistic proofs of closed formulas:

Operation 1' - the elimination of logical closure sets from the beginning part, i.e. the elimination of closure sets, 1) $\{A, \overline{A'}\}$, where A and A' are equivalent atomic formulas and 2) $\{\overline{s = t}, A, \overline{A'}\}$, where A and A' are atomic formulas equivalent relative to $s = t$.

Let us see that this can be done, assuming that operation 1 is not applicable. We first reduce the second case to the first. Since all parameters introduced in the beginning part have been eliminated, the $\overline{s = t}$ of a closure set $\{\overline{s = t}, A, \overline{A'}\}$ of the second kind must be a numerical formula. If it is false it is a (so-called "mathematical") closure set itself, being an arithmetically unsatisfiable formula. If it is true, then $\{A, \overline{A'}\}$ forms a closure set of the first kind. Now let us consider a closure set $\{A, \overline{A'}\}$ of the first kind that is being used to close a branch of the beginning part. Since only cuts and replacements can occur in the beginning part (in this particular case, we have to speak about replacements), the closure set must occur in the following way:

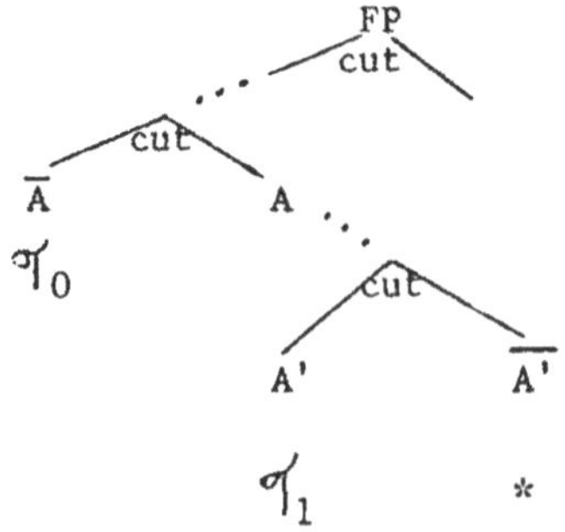

We replace this by the tableau (of lower rank):

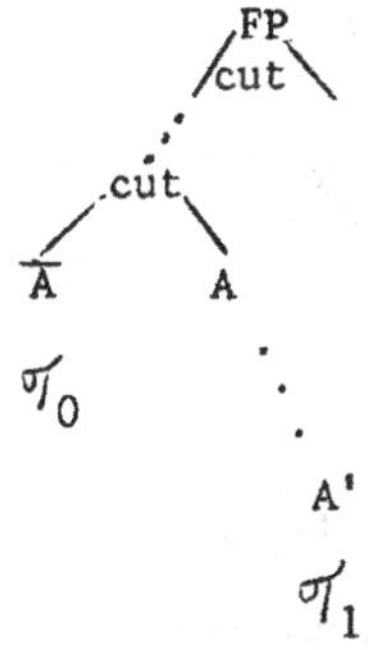

by replacement from A

We note that in a proof to which operations 1 and 1' are no longer applicable, branches of the beginning part can be closed only through the occurrence of a single false numerical formula. Let us call such a formula a mathematical falsity.

Using these results, we obtain the following proposition that will be used to obtain the main results:

<u>Lemma 1.</u> Let $\mathcal{P}$ be a (classical or intuitionistic) proof of a closed formula A such that 1) none of the Gentzen operations are applicable to $\mathcal{P}$; 2) $\mathcal{P}$ does not coincide with its beginning part. Then one of the branches of the beginning part ends with an application of an α, β, γ, or δ rule to the initial formula FA.

Proof. To obtain our result, we prove by induction on the structure of the beginning part of $\mathcal{P}$: if a formula X in the beginning part of $\mathcal{P}$ is not a mathematical falsity closing a branch, then either X is itself the premise of a logical rule, or it has such a formula above it in the proof. This clearly holds for all the final formulas of the beginning part of the proof. Now let X be the premise of some cut

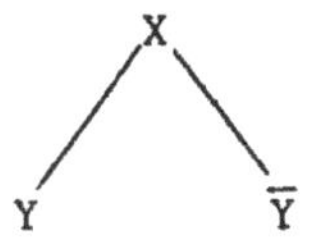

in the beginning part such that the hypothesis holds for Y and $\overline{Y}$. If one of Y, $\overline{Y}$, say Y, is a mathematical falsity, the other, $\overline{Y}$, then cannot be. But then $\overline{Y}$ is atomic, so it cannot be itself the premise of a logical rule, and the formula above it that must be the premise of a logical rule is either X or above X, as is required. On the other hand, if neither Y nor $\overline{Y}$ is an axiom, they cannot <u>both</u> be premises of logical rules (or operation 3 could be carried out on $\mathcal{P}$), so for at least one of them, the inductively given premise of the logical rule must lie above it.

We apply this proposition immediately to obtain the following main results for first order intuitionistic number theory:

<u>Theorem 54.</u> Let $\mathcal{P}$ be a proof of $(\exists x)A(x)$ in first order intuitionistic number theory, where $(\exists x)A(x)$ is a closed formula. Then a constant term t and an intuitionistic proof $\mathcal{P}'$ of A(t) can be found.

<u>Theorem 55.</u> Let $\mathcal{P}$ be a proof of $A \vee B$ in first order intuitionistic number theory, where A and B are closed formulas. Then we can effectively find an intuitionistic proof $\mathcal{P}'$ of A or of B.

Proof. We give the proof of theorem 54, the one of theorem 55 being similar.

By our previous results we can find a proof $\mathcal{P}^*$ of $(\exists x)A(x)$ upon which no operation can be carried out. Since it can be shown that in our system the only formulas that can be proved using the cut rule alone are true atomic formulas, the proof $\mathcal{P}^*$ of $(\exists x)A(x)$ cannot coincide with its beginning part. Thus by lemma 1, one of the branches of the beginning part must end with a γ rule application:

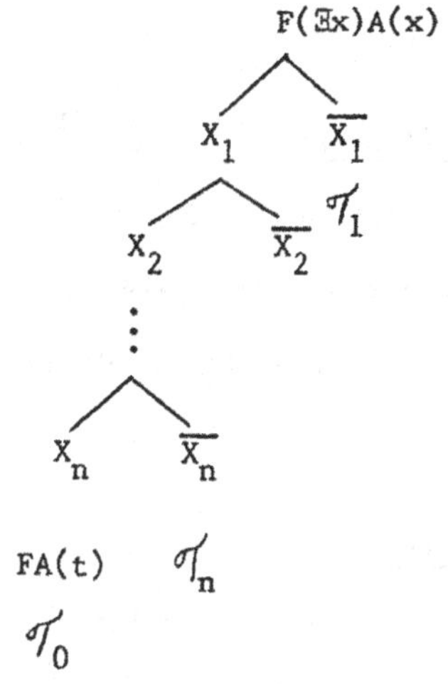

Because this is an intuitionistic proof moreover, we note that all the $X_1,\ldots,X_n$ must have the prefix T and consequently all the $\overline{X}_1,\ldots,\overline{X}_n$, the prefix F. It follows that $F(\exists x)A(x)$ cannot be used in $\mathcal{T}_1,\ldots,\mathcal{T}_n$, and thus that the following is an intuitionistic proof of $A(t)$:

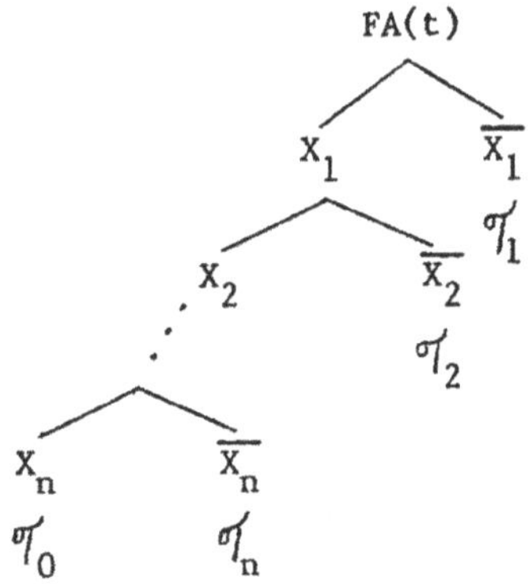

We note that these techniques can also be used to obtain (cf. Scarpellini [1]):

<u>Theorem 56 (Ackermann).</u> Let $\mathcal{P}$ be a classical proof of $(\exists x)p(x)$, where $(\exists x)p(x)$ is a closed formula. Then one can effectively find certain numerals $n_1, \ldots, n_s$ and a proof $\mathcal{P}'$ of $p(n_1) \vee \cdots \vee p(n_s)$ such that $\mathcal{P}'$ coincides with its beginning part.

<u>Theorem 57 (Harrop).</u> Let M be the class of formulas defined as follows: a) all atomic formulas are in M; b) if A and B are in M, then so is $A \wedge B$; c) if A is in M, so is $(\forall x)A$; d) if A is in M, and B is arbitrary, $B \supset A$ is in M. Now let A and $(\exists x)B(x)$ be closed formulas with $A \in M$. If there is an intuitionistic proof of $A \supset (\exists x)B(x)$, then a constant term t and an intuitionistic proof of $A \supset B(t)$ can be found.

The results we have discussed can also be easily used (through a separate consideration of the cases in which the different Gentzen operations are applicable and use of transfinite induction on the rank of the proof) to show:

<u>Theorem 58.</u> One can effectively transform a proof $\mathcal{P}$ in Ψ of rank α into a <u>cut-free</u> proof $\mathcal{P}'$ in Ω (of the same formula) with rank $\beta \leq \alpha$. If $\mathcal{P}$ is intuitionistic, so is $\mathcal{P}'$.

In [3] Scarpellini was able to use nonconstructive reasoning to extend the methods we have been discussing to obtain theorems similar to theorems 54 to 58 for certain theories of intuitionistic analysis. It had been known that, due to a result of Kreisel, these techniques could not be applied to sufficiently strong systems of classical analysis. Scarpellini noted, however, that a very simple property of intuitionistic formalisms changed the situation in the intuitionistic case. This property is the following:

<u>Lemma 2 (Scarpellini's "Basic Lemma").</u> Let $\mathcal{P}$ be an intuitionistic proof of a formula A. Let A_i $(i = 1,\ldots,n)$ be the bottom-most F-prefixed formulas in the beginning part of $\mathcal{P}$, numbered from left to right according to the branch of the beginning part they occur on (where we assume that cuts branch as follows):

.)

Let B be any T-prefixed formula in the beginning part. Then

there is an intuitionistic proof of every A_i $(i = 1,\ldots,n)$ and of B whose rank is smaller than the rank of $\mathcal{P}$ except in the case of A_n.

Proof. Thus the proof $\mathcal{P}$ looks like, say,

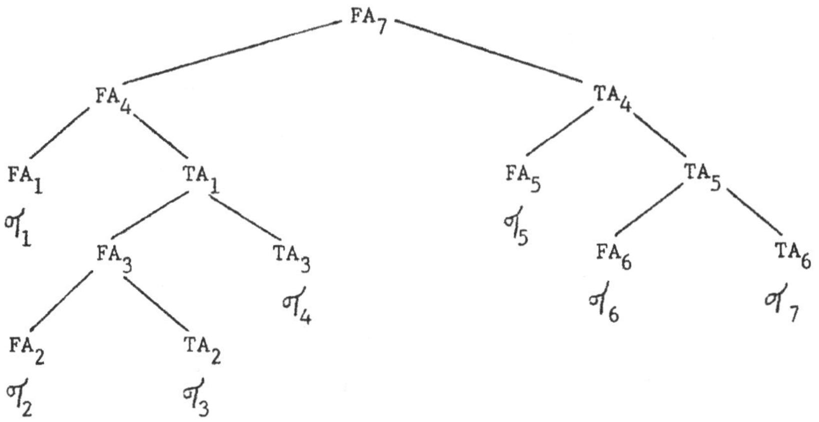

(here the branch of the beginning part ending in FA_1 is the left-most branch, the one ending in FA_2 the next left-most branch, then the branches ending in TA_2, TA_3, FA_5, FA_6, TA_6; thus FA_i is the lowest F-prefixed formula on the i-th branch from the left), where the trees $\mathcal{T}_1,\ldots,\mathcal{T}_7$ form the part of the tableau below the beginning part. Let us prove by induction that each A_i has a proof. (We will ignore the question of ordinals, which is not difficult to see, however.) This is true for A_1, since $\mathcal{T}_1$ is a proof of A_1. Now assume it is true for all A_j, $j < i$. If we let $B_1,\ldots,B_m$ be all the T-prefixed formulas on the path from FA_i to the top formula FA (here FA_7), and let $\mathcal{T}^i$ be the tree below FA_i, then

$$\begin{array}{c} TB_1 \\ \vdots \\ TB_m \\ FA_i \\ \mathcal{T}^i \end{array}$$

is a closed intuitionistic tableau for the set of formulas $\{TB_1,\ldots,TB_m,FA_i\}$. But it can be easily seen that in $\mathcal{P}$ each of the TB_k is the right cut formula of a cut whose left cut formula is a TA_j with $j < i$. Thus FB_k (i.e. FA_j) has a closed tableau

$$\begin{array}{c} FB_k \\ \mathcal{S}_k \end{array}$$

by the induction hypothesis, and we can form a proof for A_i:

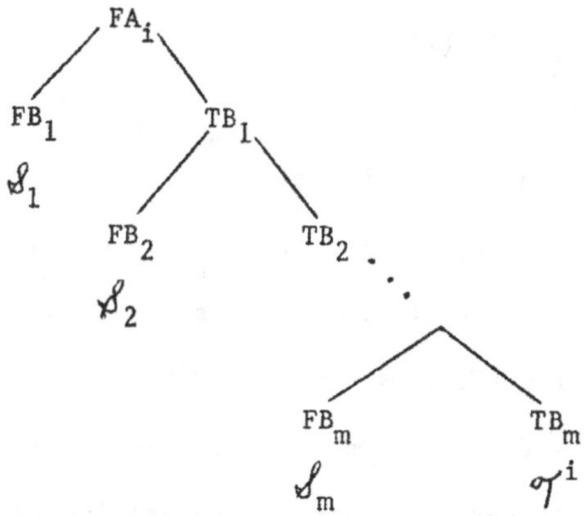

Now the second part of the lemma, namely that if TB occurs in the beginning part of $\mathcal{P}$, a proof of B can be found, can be easily seen to follow from the first part. For if TB occurs in the beginning part of the proof, it must occur as the right cut formula of a cut. Then FB is the left cut formula of the cut, and is the bottom-most F-prefixed formula on the branch that goes down from it, if one always goes in the T direction at a cut. Thus by the first part of the lemma, an intuitionistic proof for B can be effectively found. This completes the proof of the lemma.

Finally, let us just see briefly an example of how Scarpellini uses t proposition. He is interested in systems of second order arithmetic extended by certain rules of transfinite induction, for example by

$$\begin{array}{l} TR(t) \\ T(\forall F)(\exists x)(\neg F(x+1) \underset{R}{\subset} f(x)) \\ FA(t) \\ \quad | \\ TR(a) \\ T(\forall x)(x \underset{R}{\subset} a \supset A(x)) \\ FA(a) \end{array}$$

where a is new on the branch. Here the second formula of the premise, which we will denote $W(\underset{R}{\subset})$, says that $\underset{R}{\subset}$ is well-founded. Gentzen-type operations are applied to these rules, also, and in the operation Scarpellini uses for this rule, he needs to know he can find a proof of $W(\underset{R}{\subset})$. This follows from his basic lemma, for one only wants to apply the reduction operation when an application of the rule in question ends a branch in the beginning part, which means that the premises of the rule have to occur in the beginning part, as required by the basic lemma.

BIBLIOGRAPHY

1. Ackermann, W. [1] Zur Widerspruchsfreiheit der Zahlentheorie. Math. Ann. 117, 162-194 (1940). [2] Widerspruchsfreier Aufbau einer typenfreien Logik. (Erweitertes System.) Math. Z. 55, 364-384 (1952).

2. Bishop, E. [1] Foundations of Constructive Analysis. McGraw-Hill, New York, 1967.

3. Feferman, S. [1] Transfinite progressions of axiomatic theories. Journ. Symbolic Logic, 27, 259-316 (1962). [2] Systems of predicative analysis. Journ. Symbolic Logic 29, 1-3 (1964). [3] Systems of predicative analysis, II: representations of ordinals. Journ. Symbolic Logic 33, 193-220 (1968). [4] Autonomous transfinite progressions and the extent of predicative mathematics, Proc. 3rd Int'l Cong. for Logic, Methodology and Philos. of Science, Amsterdam (1967).

4. Gentzen, G. [1] Die Widerspruchsfreiheit der reinen Zahlentheorie. Math. Ann. 112, 493-565 (1936). [2] Neue Fassung des Widerspruchsfreiheitsbeweises für die reine Zahlentheorie. Forschungen zur Logik und zur Grundlegung der Exakten Wissenschaften, Nueu Folge 4, 19-44 (1938). [3] Beweisbarkeit und Unbeweisbarkeit von Anfangsfällen der transfiniten Induktion in der reinen Zahltheorie. Math. Ann. 119, 140-161 (1943). These three papers have been translated into English (entitled The consistency of elementary number theory (132-213), New version of the consistency proof for elementary number theory (252-286), and Provability and nonprovability of restricted transfinite induction in elementary number theory (287-308), respectively) and appear in Szabo, M., The Collected Papers of Gerhard Gentzen, North-Holland Press, Amsterdam, 1969.

5. Girard, J. Y. [1] Une extension de l'interpretation de Gödel à l'analyse, et son application a l'élimination des coupures dan l'analyse et la théorie des types. Proc. 2nd Scandinavian Logic Symposium, 63-92, Amsterdam (1971).

6. Gödel, K. [1] Die Vollständigkeit der Axiome des logischen Funktionenkalkuls. Monatshefte für Mathematik und Physik 37, 349-360 (1930). [2] Über formal unentscheidbare Sätze der Principia mathematica und verwandter Systeme I. Monatshefte für Mathematik und Physik 38, 173-198 (1931). [3] Über Vollständigkeit und Widerspruchsfreiheit. Ergebnisse eines Mathematischen Kolloquiums 3, 12-13 (1932). These three papers are translated into English (entitled The completeness of the axioms of the functional calculus of logic (582-591), On formally undecidable propositions of Principia Mathematica and related systems (596-616), and On completeness and consistency (616-617), respectively) and appear in van Heijenoort, J., <u>From Frege to Gödel</u>, Harvard University Press, Cambridge, 1967. [4] Zur intuitionistischen Arithmetik und Zahlentheorie. Ergebnisse eines Mathematischen Kolloquiums 4, 34-38 (for 1931-32, published 1933). [5] Über eine bisher noch nicht benutzte Erweiterung des finiten Standpunktes. Dialectica 12, 280-287 (1958).

7. Henkin, L. [1] Completeness in the theory of types, Journ. Symbolic Logic 15, 81-91 (1950).

8. Hilbert, D. [1] Über das unendlichen. Math. Ann. 95, 161-190 (1926). [2] Die Grundlagen der Mathematik. Abhandlungen aus dem mathematischen Seminar der Hamburgischen Universität 6, 65-85 (1928). These papers are translated into English (On the infinite (367-392) and The foundations of mathematics (464-479) and appear on the pages referred to in van Heijenoort, J.,

From Frege to Gödel, Harvard University Press, Cambridge, 1967. [3] Probleme der Grundlegung der Mathematik. Reprinted with changes and additions in Mathematische Annalen 102, 1-9, (1929).

9. Hilbert, D. and Bernays, P. [1] Grundlagen der Mathematik, Springer-Verlag, Berlin, 1939.

10. Hintikka, J. [1] Form and content in quantification theory. Two papers on symbolic logic. Acta Philosophica Fennica 8, 7-55, (1955).

11. Howard, W. [1] Assignment of ordinals to terms for primitive recursive arithmetic. Intuitionism and Proof Theory (Proc. of Summer Conf., Buffalo, 1968), 443-458, North-Holland Press, Amsterdam, (1970).

12. Kleene, S. [1] Introduction to Metamathematics. Van Nostrand, Princeton, 1952.

13. Kreisel, G. [1] A survey of proof theory. Journ. Symbolic Logic 33, 321-388 (1968). [2] A survey of proof theory, II. Proc. 2nd Scandinavian Logic Symposium, 109-170, North-Holland Press, Amsterdam (1971).

14. Kreisel, G., Schoenfield, J., and Wang, H. [1] Number theoretic concepts and recursive well-orderings. Archiv für mathematische Logik und Grundlagenforschung 5, 42-64 (1960).

15. Luckhardt, H. [1] Extensional Gödel Functional Interpretation: A consistency proof for classical analysis. Lecture Notes in Mathematics No. 306, Springer-Verlag, Berlin, 1973.

16. Martin-Löf, P. [1] Hauptsatz for the theory of species. Proc. 2nd Scandinavian Logic Symposium, 217-233, North-Holland Press, Amsterdam (1971).

17. Myhill, J. [1] A stumbling block in constructive mathematics. Journ. Symbolic Logic 18, 190-191 (1953).

18. Prawitz, D. [1] Completeness and Hauptsatz for second order logic. Theoria 33, 246-258 (1967). [2] Hauptsatz for higher order logic. Journ. Symbolic Logic 33, 452-457 (1968).

19. Scarpellini, B. [1] Some applications of Gentzen's second consistency proof. Math. Ann. 181, 325-344 (1969). [2] On cut elimination in intuitionistic systems of analysis. <u>Intuitionism and Proof Theory</u> (Proc. of Summer Conf., Buffalo, 1968), North-Holland Press, Amsterdam (1970). [3] <u>Proof Theory and Intuitionistic Systems</u>. Lecture Notes in Mathematics No. 212, Springer-Verlag, Berlin, 1971.

20. Schütte, K. [1] Beweistheoretische Erfassung der unendlichen Induktion in der Zahlentheorie. Math. Ann. 122, 369-389 (1951). [2] Beweistheoretische Untersuchung der verzweigten Analysis. Math. Ann. 124, 123-147 (1952). [3] Zur Widerspruchsfreiheit einer typenfreien Logik. Math. Ann. 125, 394-400 (1953). [4] Kennzeichnung von Ordnungszahlen durch rekursiv erklärte Funktionen. Math. Ann. 127, 15-32 (1954). [5] <u>Beweistheorie</u>. Springer, Berlin, 1960. [6] Syntactical and semantical properties of simple type theory. Journ. Symbolic Logic 25, 305-326 (1960). [7] <u>Lecture Notes in Metamathematics</u>. The Pennsylvania State University, 1962.

21. Smullyan, R. [1] <u>First-Order Logic</u>. Springer, Berlin, 1968. [2] Abstract quantification theory. <u>Intuitionism and Proof Theory</u> (Proc. of Summer Conf., Buffalo, 1968), North-Holland Press, Amsterdam, (1970).

22. Stenius, E. [1] Das Interpretationsproblem der formalisierten Zahlentheorie und ihre formale Widerspruchsfreiheit. Acta Acad. Abrensis, Math. Phys. 18, Nr. 3 (1952).

23. Tait, W. [1] A nonconstructive proof of Gentzen's Hauptsatz for second order predicate logic. Bull. Amer. Math. Soc. 72, 980-983 (1966). [2] Intensional interpretation of functionals of finite type. Journ. Symbolic Logic 32, 198-212 (1967).

24. Takahashi, M. [1] A proof of cut-elimination theorem in simple type theory. J. Math. Soc. Japan 19, 399-410 (1967).

25. Takeuti, G. [1] On a generalized logical calculus. Japan J. Math. 23, 39-96 (1953). [2] On the fundamental conjecture of GLC I. J. Math. Soc. Japan 7, 249-275 (1955). [3] Consistency proofs of subsystems of classical analysis. Ann. Math. (2), 86, 299-348 (1967). [4] Π_1^1-comprehension axioms and ω-rule. Proc. Institute in Logic, Leeds (1967). [5] A conservative extension of Peano's arithmetic. To appear.

26. Turing, A. [1] On computable numbers, with an application to the entscheidungsproblem. Proc. London Math. Soc. 42, 230-265 (1963-7). [2] Systems of logic based on ordinals. Proc. London Math. Soc. 45, 161-228 (1939). Both these papers are reprinted in Davis, M., The Undecidable, Raven Press, Hewlett, 1965.

27. Wang, H. [1] From Mathematics to Philosophy. Routledge and Kegan Paul, London, 1974.

28. Yasuhara, M. [1] A constructive version of Hilbert's Nullstellensatz. Private memorandum.

ol. 277: Séminaire Banach. Edité par C. Houzel. VII, 229 pages. 972. DM 22,-

ol. 278: H. Jacquet, Automorphic Forms on GL(2) Part II. XIII, 142 ages. 1972. DM 18,-

ol. 279: R. Bott, S. Gitler and I. M. James, Lectures on Algebraic and fferential Topology. V, 174 pages. 1972. DM 20,-

ol. 280: Conference on the Theory of Ordinary and Partial Differenal Equations. Edited by W. N. Everitt and B. D. Sleeman. XV, 367 ages. 1972. DM 29,-

ol. 281: Coherence in Categories. Edited by S. Mac Lane. VII, 235 ages. 1972. DM 22,-

ol. 282: W. Klingenberg und P. Flaschel, Riemannsche Hilbertmangfaltigkeiten. Periodische Geodätische. VII, 211 Seiten. 1972. M 22,-

ol. 283: L. Illusie, Complexe Cotangent et Déformations II. VII, 304 ages. 1972. DM 27,-

ol. 284: P. A. Meyer, Martingales and Stochastic Integrals I. VI, 89 ages. 1972. DM 18,-

ol. 285: P. de la Harpe, Classical Banach-Lie Algebras and Banache Groups of Operators in Hilbert Space. III, 160 pages. 1972. M 18,-

ol. 286: S. Murakami, On Automorphisms of Siegel Domains. V, 5 pages. 1972. DM 18,-

l. 287: Hyperfunctions and Pseudo-Differential Equations. Edited H. Komatsu. VII, 529 pages. 1973. DM 40,-

l. 288: Groupes de Monodromie en Géométrie Algébrique. (SGA 7 Dirigé par A. Grothendieck. IX, 523 pages. 1972. DM 55,-

l. 289: B. Fuglede, Finely Harmonic Functions. III, 188. 1972. M 20,-

l. 290: D. B. Zagier, Equivariant Pontrjagin Classes and Applicans to Orbit Spaces. IX, 130 pages. 1972. DM 18,-

l. 291: P. Orlik, Seifert Manifolds. VIII, 155 pages. 1972. DM 18,-

l. 292: W. D. Wallis, A. P. Street and J. S. Wallis, Combinatorics: om Squares, Sum-Free Sets, Hadamard Matrices. V, 508 pages. 72. DM 55,-

l. 293: R. A. DeVore, The Approximation of Continuous Functions Positive Linear Operators. VIII, 289 pages. 1972. DM 27,-

l. 294: Stability of Stochastic Dynamical Systems. Edited by R. F. urtain. IX, 332 pages. 1972. DM 29,-

l. 295: C. Dellacherie, Ensembles Analytiques Capacités. Mesures Hausdorff. XII, 123 pages. 1972. DM 18,-

l. 296: Probability and Information Theory II. Edited by M. Behara, Krickeberg and J. Wolfowitz. V, 223 pages. 1973. DM 22,-

l. 297: J. Garnett, Analytic Capacity and Measure. IV, 138 pages. 72. DM 18,-

l. 298: Proceedings of the Second Conference on Compact Transrmation Groups. Part 1. XIII, 453 pages. 1972. DM 35,-

l. 299: Proceedings of the Second Conference on Compact Transrmation Groups. Part 2. XIV, 327 pages. 1972. DM 29,-

l. 300: P. Eymard, Moyennes Invariantes et Représentations Unires. II, 113 pages. 1972. DM 18,-

l. 301: F. Pittnauer, Vorlesungen über asymptotische Reihen. VI, 6 Seiten. 1972. DM 18,-

l. 302: M. Demazure, Lectures on p-Divisible Groups. V, 98 pages. 72. DM 18,-

l. 303: Graph Theory and Applications. Edited by Y. Alavi, D. R. Lick d A. T. White. IX, 329 pages. 1972. DM 26,-

l. 304: A. K. Bousfield and D. M. Kan, Homotopy Limits, Compleons and Localizations. V, 348 pages. 1972. DM 29,-

l. 305: Théorie des Topos et Cohomologie Etale des Schémas. me 3. (SGA 4). Dirigé par M. Artin, A. Grothendieck et J. L. Verdier. 640 pages. 1973. DM 55,-

l. 306: H. Luckhardt, Extensional Gödel Functional Interpretation. 161 pages. 1973. DM 20,-

l. 307: J. L. Bretagnolle, S. D. Chatterji et P.-A. Meyer, Ecole d'été Probabilités: Processus Stochastiques. VI, 198 pages. 1973. M 22,-

l. 308: D. Knutson, λ-Rings and the Representation Theory of the mmetric Group. IV, 203 pages. 1973. DM 22,-

Vol. 309: D. H. Sattinger, Topics in Stability and Bifurcation Theory. VI, 190 pages. 1973. DM 20,-

Vol. 310: B. Iversen, Generic Local Structure of the Morphisms in Commutative Algebra. IV, 108 pages. 1973. DM 18,-

Vol. 311: Conference on Commutative Algebra. Edited by J. W. Brewer and E. A. Rutter. VII, 251 pages. 1973. DM 24,-

Vol. 312: Symposium on Ordinary Differential Equations. Edited by W. A. Harris, Jr. and Y. Sibuya. VIII, 204 pages. 1973. DM 22,-

Vol. 313: K. Jörgens and J. Weidmann, Spectral Properties of Hamiltonian Operators. III, 140 pages. 1973. DM 18,-

Vol. 314: M. Deuring, Lectures on the Theory of Algebraic Functions of One Variable. VI, 151 pages. 1973. DM 18,-

Vol. 315: K. Bichteler, Integration Theory (with Special Attention to Vector Measures). VI, 357 pages. 1973. DM 29,-

Vol. 316: Symposium on Non-Well-Posed Problems and Logarithmic Convexity. Edited by R. J. Knops. V, 176 pages. 1973. DM 20,-

Vol. 317: Séminaire Bourbaki – vol. 1971/72. Exposés 400–417. IV, 361 pages. 1973. DM 29,-

Vol. 318: Recent Advances in Topological Dynamics. Edited by A. Beck. VIII, 285 pages. 1973. DM 27,-

Vol. 319: Conference on Group Theory. Edited by R. W. Gatterdam and K. W. Weston. V, 188 pages. 1973. DM 20,-

Vol. 320: Modular Functions of One Variable I. Edited by W. Kuyk. V, 195 pages. 1973. DM 20,-

Vol. 321: Séminaire de Probabilités VII. Edité par P. A. Meyer. VI, 322 pages. 1973. DM 29,-

Vol. 322: Nonlinear Problems in the Physical Sciences and Biology. Edited by I. Stakgold, D. D. Joseph and D. H. Sattinger. VIII, 357 pages. 1973. DM 29,-

Vol. 323: J. L. Lions, Perturbations Singulières dans les Problèmes aux Limites et en Contrôle Optimal. XII, 645 pages. 1973. DM 46,-

Vol. 324: K. Kreith, Oscillation Theory. VI, 109 pages. 1973. DM 18,-

Vol. 325: C.-C. Chou, La Transformation de Fourier Complexe et L'Equation de Convolution. IX, 137 pages. 1973. DM 18,-

Vol. 326: A. Robert, Elliptic Curves. VIII, 264 pages. 1973. DM 24,-

Vol. 327: E. Matlis, One-Dimensional Cohen-Macaulay Rings. XII, 157 pages. 1973. DM 20,-

Vol. 328: J. R. Büchi and D. Siefkes, The Monadic Second Order Theory of All Countable Ordinals. VI, 217 pages. 1973. DM 22,-

Vol. 329: W. Trebels, Multipliers for (C, α)-Bounded Fourier Expansions in Banach Spaces and Approximation Theory. VII, 103 pages. 1973. DM 18,-

Vol. 330: Proceedings of the Second Japan-USSR Symposium on Probability Theory. Edited by G. Maruyama and Yu. V. Prokhorov. VI, 550 pages. 1973. DM 40,-

Vol. 331: Summer School on Topological Vector Spaces. Edited by L. Waelbroeck. VI, 226 pages. 1973. DM 22,-

Vol. 332: Séminaire Pierre Lelong (Analyse) Année 1971-1972. V, 131 pages. 1973. DM 18,-

Vol. 333: Numerische, insbesondere approximationstheoretische Behandlung von Funktionalgleichungen. Herausgegeben von R. Ansorge und W. Törnig. VI, 296 Seiten. 1973. DM 27,-

Vol. 334: F. Schweiger, The Metrical Theory of Jacobi-Perron Algorithm. V, 111 pages. 1973. DM 18,-

Vol. 335: H. Huck, R. Roitzsch, U. Simon, W. Vortisch, R. Walden, B. Wegner und W. Wendland, Beweismethoden der Differentialgeometrie im Großen. IX, 159 Seiten. 1973. DM 20,-

Vol. 336: L'Analyse Harmonique dans le Domaine Complexe. Edité par E. J. Akutowicz. VIII, 169 pages. 1973. DM 20,-

Vol. 337: Cambridge Summer School in Mathematical Logic. Edited by A. R. D. Mathias and H. Rogers. IX, 660 pages. 1973. DM 46,-

Vol: 338: J. Lindenstrauss and L. Tzafriri, Classical Banach Spaces. IX, 243 pages. 1973. DM 24,-

Vol. 339: G. Kempf, F. Knudsen, D. Mumford and B. Saint-Donat, Toroidal Embeddings I. VIII, 209 pages. 1973. DM 22,-

Vol. 340: Groupes de Monodromie en Géométrie Algébrique. (SGA 7 II). Par P. Deligne et N. Katz. X, 438 pages. 1973. DM 44,-

Vol. 341: Algebraic K-Theory I, Higher K-Theories. Edited by H. Bass. XV, 335 pages. 1973. DM 29,-

Vol. 342: Algebraic K-Theory II, "Classical" Algebraic K-Theory, and Connections with Arithmetic. Edited by H. Bass. XV, 527 pages. 1973. DM 40,-

Vol. 343: Algebraic K-Theory III, Hermitian K-Theory and Geometric Applications. Edited by H. Bass. XV, 572 pages. 1973. DM 40,-

Vol. 344: A. S. Troelstra (Editor), Metamathematical Investigation of Intuitionistic Arithmetic and Analysis. XVII, 485 pages. 1973. DM 38,-

Vol. 345: Proceedings of a Conference on Operator Theory. Edited by P. A. Fillmore. VI, 228 pages. 1973. DM 22,-

Vol. 346: Fučík et al., Spectral Analysis of Nonlinear Operators. II, 287 pages. 1973. DM 26,-

Vol. 347: J. M. Boardman and R. M. Vogt, Homotopy Invariant Algebraic Structures on Topological Spaces. X, 257 pages. 1973. DM 24,-

Vol. 348: A. M. Mathai and R. K. Saxena, Generalized Hypergeometric Functions with Applications in Statistics and Physical Sciences. VII, 314 pages. 1973. DM 26,-

Vol. 349: Modular Functions of One Variable II. Edited by W. Kuyk and P. Deligne. V, 598 pages. 1973. DM 38,-

Vol. 350: Modular Functions of One Variable III. Edited by W. Kuyk and J.-P. Serre. V, 350 pages. 1973. DM 26,-

Vol. 351: H. Tachikawa, Quasi-Frobenius Rings and Generalizations. XI, 172 pages. 1973. DM 20,-

Vol. 352: J. D. Fay, Theta Functions on Riemann Surfaces. V, 137 pages. 1973. DM 18,-

Vol. 353: Proceedings of the Conference. on Orders, Group Rings and Related Topics. Organized by J. S. Hsia, M. L. Madan and T. G. Ralley. X, 224 pages. 1973. DM 22,-

Vol. 354: K. J. Devlin, Aspects of Constructibility. XII, 240 pages. 1973. DM 24,-

Vol. 355: M. Sion, A Theory of Semigroup Valued Measures. V, 140 pages. 1973. DM 18,-

Vol. 356: W. L. J. van der Kallen, Infinitesimally Central-Extensions of Chevalley Groups. VII, 147 pages. 1973. DM 18,-

Vol. 357: W. Borho, P. Gabriel und R. Rentschler, Primideale in Einhüllenden auflösbarer Lie-Algebren. V, 182 Seiten. 1973. DM 20,-

Vol. 358: F. L. Williams, Tensor Products of Principal Series Representations. VI, 132 pages. 1973. DM 18,-

Vol. 359: U. Stammbach, Homology in Group Theory. VIII, 183 pages. 1973. DM 20,-

Vol. 360: W. J. Padgett and R. L. Taylor, Laws of Large Numbers for Normed Linear Spaces and Certain Fréchet Spaces. VI, 111 pages. 1973. DM 18,-

Vol. 361: J. W. Schutz, Foundations of Special Relativity: Kinematic Axioms for Minkowski Space Time. XX, 314 pages. 1973. DM 26,-

Vol. 362: Proceedings of the Conference on Numerical Solution of Ordinary Differential Equations. Edited by D. Bettis. VIII, 490 pages. 1974. DM 34,-

Vol. 363: Conference on the Numerical Solution of Differential Equations. Edited by G. A. Watson. IX, 221 pages. 1974. DM 20,-

Vol. 364: Proceedings on Infinite Dimensional Holomorphy. Edited by T. L. Hayden and T. J. Suffridge. VII, 212 pages. 1974. DM 20,-

Vol. 365: R. P. Gilbert, Constructive Methods for Elliptic Equations. VII, 397 pages. 1974. DM 26,-

Vol. 366: R. Steinberg, Conjugacy Classes in Algebraic Groups (Notes by V. V. Deodhar). VI, 159 pages. 1974. DM 18,-

Vol. 367: K. Langmann und W. Lütkebohmert, Cousinverteilungen und Fortsetzungssätze. VI, 151 Seiten. 1974. DM 16,-

Vol. 368: R. J. Milgram, Unstable Homotopy from the Stable Point of View. V, 109 pages. 1974. DM 16,-

Vol. 369: Victoria Symposium on Nonstandard Analysis. Edited by A. Hurd and P. Loeb. XVIII, 339 pages. 1974. DM 26,-

Vol. 370: B. Mazur and W. Messing, Universal Extensions an One Dimensional Crystalline Cohomology. VII, 134 pages. 1974 DM 16,-

Vol. 371: V. Poenaru, Analyse Différentielle. V, 228 pages. 1974 DM 20,-

Vol. 372: Proceedings of the Second International Conferenc on the Theory of Groups 1973. Edited by M. F. Newman. VI 740 pages. 1974. DM 48,-

Vol. 373: A. E. R. Woodcock and T. Poston, A Geometrical Stud of the Elementary Catastrophes. V, 257 pages. 1974. DM 22,

Vol. 374: S. Yamamuro, Differential Calculus in Topologic Linear Spaces. IV, 179 pages. 1974. DM 18,-

Vol. 375: Topology Conference 1973. Edited by R. F. Dickma Jr. and P. Fletcher. X, 283 pages. 1974. DM 24,-

Vol. 376: D. B. Osteyee and I. J. Good, Information, Weight o Evidence, the Singularity between Probability Measures an Signal Detection. XI, 156 pages. 1974. DM 16.-

Vol. 377: A. M. Fink, Almost Periodic Differential Equation VIII, 336 pages. 1974. DM 26,-

Vol. 378: TOPO 72 - General Topology and its Application Proceedings 1972. Edited by R. Alò, R. W. Heath and J. Nagat XIV, 651 pages. 1974. DM 50,-

Vol. 379: A. Badrikian et S. Chevet, Mesures Cylindrique Espaces de Wiener et Fonctions Aléatoires Gaussiennes. X 383 pages. 1974. DM 32,-

Vol. 380: M. Petrich, Rings and Semigroups. VIII, 182 page 1974. DM 18,-

Vol. 381: Séminaire de Probabilités VIII. Edité par P. A. Meye IX, 354 pages. 1974. DM 32,-

Vol. 382: J. H. van Lint, Combinatorial Theory Seminar Ein hoven University of Technology. VI, 131 pages. 1974. DM 18,

Vol. 383: Séminaire Bourbaki - vol. 1972/73. Exposés 418-43 IV, 334 pages. 1974. DM 30,-

Vol. 384: Functional Analysis and Applications, Proceeding 1972. Edited by L. Nachbin. V, 270 pages. 1974. DM 22,-

Vol. 385: J. Douglas Jr. and T. Dupont, Collocation Methods fc Parabolic Equations in a Single Space Variable (Based on C Piecewise-Polynomial Spaces). V, 147 pages. 1974. DM 16,

Vol. 386: J. Tits, Buildings of Spherical Type and Finite BN Pairs. IX, 299 pages. 1974. DM 24,-

Vol. 387: C. P. Bruter, Eléments de la Théorie des Matroïde V, 138 pages. 1974. DM 18,-

Vol. 388: R. L. Lipsman, Group Representations. X, 166 page 1974. DM 20,-

Vol. 389: M.-A. Knus et M. Ojanguren, Théorie de la Descent et Algèbres d' Azumaya. IV, 163 pages. 1974. DM 20,-

Vol. 390: P. A. Meyer, P. Priouret et F. Spitzer, Ecole d'Eté c Probabilités de Saint-Flour III - 1973. Edité par A. Badrikia et P.-L. Hennequin. VIII, 189 pages. 1974. DM 20,-

Vol. 391: J. Gray, Formal Category Theory: Adjointness for Categories. XII, 282 pages. 1974. DM 24,-

Vol. 392: Géométrie Différentielle, Colloque, Santiago c Compostela, Espagne 1972. Edité par E. Vidal. VI, 225 page 1974. DM 20,-

Vol. 393: G. Wassermann, Stability of Unfoldings. IX, 164 page 1974. DM 20,-

Vol. 394: W. M. Patterson 3rd. Iterative Methods for the Solutic of a Linear Operator Equation in Hilbert Space - A Surve III, 183 pages. 1974. DM 20,-

Vol. 395: Numerische Behandlung nichtlinearer Integrodifferer tial- und Differentialgleichungen. Tagung 1973. Herausgegebe von R. Ansorge und W. Törnig. VII, 313 Seiten. 1974. DM 28

Vol. 396: K. H. Hofmann, M. Mislove and A. Stralka, The Pontr agin Duality of Compact O-Dimensional Semilattices and i Applications. XVI, 122 pages. 1974. DM 18,-

Vol. 397: T. Yamada, The Schur Subgroup of the Brauer Grou V, 159 pages. 1974. DM 18,-

Vol. 398: Théories de l'Information, Actes des Rencontres d Marseille-Luminy, 1973. Edité par J. Kampé de Fériet et C. Picar XII, 201 pages. 1974. DM 23,-